SAN LUIS OBISPO COUNTY
WINE

SAN LUIS OBISPO COUNTY WINE

A WORLD-CLASS HISTORY

LIBBIE AGRAN AND HEATHER MURAN

with the Wine History Project of San Luis Obispo County

Published by American Palate
A Division of The History Press
Charleston, SC
www.historypress.com

Front cover: (*Top half*) Edna Valley in the 1960s. *Courtesy of San Luis Obispo Coast Wine.* (*Bottom half*) Paragon Vineyards in the Edna Valley at sunrise. *Courtesy of Barry Goyette.* (*Inset left*) Bill York on a tractor in a York Mountain vineyard. *Courtesy of Jan York.* (*Inset right*) Grape stomping to crush grapes has been used for thousands of years to make wine and celebrate the end of the harvest. These children, seen here stomping grapes, are celebrating home winemaking during the Prohibition era. *Courtesy of the Wine History Project of San Luis Obispo County.*

Back cover: (*Bottom half*) Maison Duetz Winery became the leader in making Méthode Champenoise sparkling wines. It was the only winery in California and the western hemisphere to operate two French Coquard presses. *Courtesy of Laetitia Winery.* (*Inset*) Zinfandel, the heritage grape of San Luis Obispo County. *Courtesy of the Wine History Project of San Luis Obispo County.*

First published 2021

Manufactured in the United States

ISBN 9781467146296

Library of Congress Control Number: 2020948629

Every glass of wine contains local history, enriching the flavor and enjoyment.

This book is dedicated to all the grape growers and winemakers who have contributed their hard work to making the unique wines of San Luis Obispo County.

In 2015, Libbie Agran established the Wine History Project to study the land, microclimates, grape varieties, growers and winemakers who have shaped the wine history of San Luis Obispo County.

Today, the Wine History Project of San Luis Obispo County is staffed by historians and museum professionals who collaborate with a diverse group of advisors and founders. It continues to document and preserve the unique wine and food history of San Luis Obispo County.

Contents

Acknowledgements

The Wine History Project of San Luis Obispo County

Aimee Armour-Avant, information designer, Wine History Project
Cynthia Lambert, curator, Wine History Project
Karen Petersen, librarian, Wine History Project
Central Coast Wine Classic Archives at the Wine History Project
Claiborne and Churchill Archives at the Wine History Project
Dusi Family Archives at the Wine History Project
Gary Eberle Archives at the Wine History Project
Max Goldman and York Mountain Winery Archives at the Wine History Project
Archie McLaren Archives at the Wine History Project
Saucelito Canyon Winery Archives at the Wine History Project
Richard Sauret Archives at the Wine History Project

Partners Archiving and Preserving Local History

El Paso de Robles Historical Society
Paso Robles Pioneer Museum
Cal Poly Kennedy Library, Special Collections
UC Davis Special Collections
Mission San Miguel de Arcángel
Mission San Luis Obispo de Tolosa

Oral Interviews

Neil and Janet Abbey
July Ackerman
Beverly Aho
John Alban
Malani Anderson
Justin Baldwin
Mark Battany
Doug and Nancy Beckett
David and Judy Breitstein
James Paul Brown
Laverne and Gilbert Buckman
Don and Gail Campbell
Jan Cannon
Dave and Marc Caparone
Vicki Carroll
Ali Rush Carscaden
Dennis Cassidy
Doug and Lavonne Casteel
Jim Clendenen
Gary Conway
Kathleen Conway
Lynn Diehl
George Donati
Hank Donatoni and Sandi Baird
Stephen and Paula Dooley
Benito Dusi
Dorothy Dusi
Janell Dusi and the J. Dusi Winery
Joni and Mike Dusi
Matthew and Ali Dusi
Michael Dusi
Chris Eberle
Gary and Marcy Eberle
Jim Efird
Joann Ernst
Manu Fiorentini
Yuroz Gevorgian
Marc Goldberg and Maggie D'Ambrosia
Barry Goyette
Bill and Nancy Greenough
Tom Greenough
Jason Haas
Robert Haas
Dan Hardesty and Kathy Marcks-Hardesty
Grey Hartley
Eric, Dave and Carmen Hickey
Michael Hoffman
Terry Hoffman
Kent Kenney
Robert Klintworth
Chris Leschinsky
Bob Lindquist
Tim Lloyd
Noreen Martin
Bruno and Debbie Martinelli
Maria Martinelli
Jenna Martinez and Epoch Wines
Jim McCormick
Archie McLaren
Ian McPhee
Mark Mozart
Melody Mullis
John Munch
Tom Myers
Robert Nadeau
Bonnie Nelson
Frank Nerelli
John R. and John H. Niven
Candice Norcross
Lei Norman
Katie O'Hara
Chuck Ortman

Gwen and Don Othman
Frank Ostini
The Paderewski Festival
Tegan Passalacqua
Steve Peck
Chris and Jeff Penick
Joel Peterson and the Paso Robles Country Wine Alliance
Suzanne Goldman Redberg
Neil Roberts
Christian Roguenant
John Rolf
Anatoly and Louise Rosinsky
Gene Sauret
Richard Sauret
Bill Scheffer
Bob Schiebelhut
Tobin James Schumrick
Larry Shupnick
Mike Sinor
James Sinton
Steve and Jane Sinton
Kirk Smith
Laura Sorvetti
Cindy Steinbeck
Howie and Bev Steinbeck
Brennen Stover
Brian and Johnine Talley and Talley Vineyards
Sue Terry
Clay and Fredericka Thompson and Claiborne and Churchill
Elizabeth Thompson
Christine Turley
Jill Tweedie
Nancy Tweedie
Niels and Bimmer Udsen and Castoro Cellars Winery
Ken Volk
Jan York
Margaret Zuech

1
The Origins of California Viticulture

Spanish Explorers and Catholic Missionaries in Alta California

GRAPE VARIETY: MISSION GRAPES (LISTÁN PRIETO)

Introduction

California viticulture history, distinct from other areas of the United States, is unique in its origins and in the first widely planted grape varieties. California viticulture originated with the Spanish Crown seeking new lands to conquer and sending explorers. In 1518, Spanish explorer Hernán Cortés sailed with his army from Cuba to conquer Mexico. He discovered Lower California (Baja) and then sent explorers north along the Pacific Coast to claim Alta California for Spain and King Charles I (also known as Holy Roman Emperor Charles V). When Cortés landed on the Pacific Coast, he decreed that each settler in Mexico and lower California had to plant one thousand grapevines for every one hundred inhabitants on his land; this marked the beginning of California viticulture.

The grapevines, *Vitis vinifera*, were brought by ship from Spain and became known in the New World as Mission grapes. The grape, almost five centuries later, has been identified as Listán Prieto, a red grape believed to have originated in the Castilla–La Mancha region of Spain. The grapevines arrived in Mexico around 1540; in the 1620s, cuttings from Mission grapes were transported north to be planted in the Spanish territory that is now recognized as the state of New Mexico.

California's viticulture is also characterized by geography, specifically the state's physical isolation from the rest of the United States. The entire west side of the narrow state has 840 miles of coastline; most of the mountain ranges run north and south, with elevations reaching 14,505 feet on the east side of the state. Santa Barbara, San Luis Obispo and Monterey Counties—regions of the Central Coast—are known for their Mediterranean-like climates. In contrast, vast deserts are located inland.

The third defining characteristic of California viticulture is its Catholic religious heritage that dates back to the mid-eighteenth century. The first chapter of California wine history was defined by twenty-one missions with chapels, lodging and vineyards established by the Spanish Franciscan padres who were led by Father Junipero Serra. The Spaniards were Catholic; the grapes and red wine were important symbols in their rituals and culture. The mission vineyards were planted by the Spanish padres and local Native Americans who were trained as agricultural workers. The Native Americans tended the drought-tolerant and hedge-pruned vines; the padres made both a dull, unstable red wine and an exciting distilled brandy, the ingredient necessary to fortify wines. Viticulture historian Charles L. Sullivan described the introduction of stills to make brandy, *aguardiente* in Spanish, as an event that changed the course of mission winemaking. The earliest reference to distilling brandy in the mission era is from 1797, according to wine historian Thomas Pinney.

EARLY SPANISH AND PORTUGUESE EXPLORERS: THE FIRST THREE HUNDRED YEARS IN CALIFORNIA

The first European explorer to visit the sand dunes in San Luis Obispo County was Juan Rodriguez Cabrillo, a Portuguese sailor who came as early as 1542. He explored the Bay of San Luis Obispo (between Point San Luis and Avila), Estero Bay, the conical rock rising in the bay known as El Moro (Morro Rock) and Piedras Blancas, where the elephant seals now breed. Cabrillo charted all of these areas and returned to San Miguel Island off the coast of Santa Barbara for the winter, where he died unexpectedly on January 5, 1543. His ships returned to Mexico, ending the first exploration and claiming of Alta California.

In 1602, the Spanish sent Sebastián Vizcaíno, a pearl fisherman, to explore Alta California. He commanded a fleet of three vessels, which slowly traveled north, surveying the Pacific Coast. Written records of his historic voyage were kept as he traveled and are now archived in Spain. It is believed that he named many modern locations in California, including San Diego, the Santa Barbara Islands, Monterey and Point Reyes.

These voyages and discoveries, however, were soon forgotten. Many Spanish kings reigned over the next 160 years, before Alta California was rediscovered. It was Charles III, who reigned from 1759 to 1788, who launched new voyages and land expeditions to survey the Pacific Coast. King Charles wanted Spain to physically occupy Alta California; he engaged the Catholic Church, a powerful religious and political force, to assist him. The religious Order of the Franciscans had built missions in Lower California to convert Native Americans to Catholicism. The Franciscan Order was chosen to continue its work in Alta California to spread Catholicism and the Spanish culture to all who resided there.

Two expeditions were organized—one by sea and one by land. The first was headed by a strong and powerful leader, José de Gálvez, who is credited with developing the idea of building missions to control Alta California. Gálvez selected Father Junipero Serra y Ferrer as his chief missionary. Gálvez's plan was to sail to three ports and found missions in San Diego, Ventura and Monterey. The second expedition was organized by land. Gaspar de Portolá, the captain of the dragoons and governor of Lower California, led the land expedition north to the first location: San Diego. Father Junipero Serra, who was appointed president of the missions, and his padres marched with Portolá and were joined by Father Juan Crespi. Crespi kept a detailed diary of the expedition, recording the valuable history of the times. The land expedition traveled with supplies, including over two hundred head of cattle and cuttings of Mission grapevines to plant in mission vineyards.

The first nine of the twenty-one missions established in Alta California were founded by Father Junipero Serra before his death in 1784. Each mission was located on a specific site that was taken in the name of the king of Spain with a religious ceremony. The missions supported the declaration of the province of New Spain in Alta California, an expansion of the Spanish Empire. Each site was commemorated with a symbolic structure, whether it was a tent, an arbor or a hanging bell on a tree branch. Construction on the sites started almost immediately; small vineyards of a few acres were planted shortly thereafter, and cattle grazed in the fields. Most of the missions, from San Diego (1769) to Sonoma (1823), were built near the road now known as

Father Junipero Serra y Ferrer, president of the missions, selected the site to build Mission San Luis Obispo de Tolosa in 1772. *Photograph courtesy of Wikimedia.*

the El Camino Real, which winds north along the Pacific Coast. You can still drive the route today and visit the missions.

The missions are among the oldest surviving structures in California. Each one is unique in its design and decoration. Many provide religious services weekly, as well as host historic collections, religious festivals and musical events. The cities of Los Angeles, San Diego, San Jose and San Francisco developed as outgrowths of these missions, which have become major tourist attractions in California.

In 1769, Don Gaspar de Portolá and his expedition explored much of the Central Coast and camped on the sand dunes, and for the first time, they saw the Valley of the Bears, where grizzly bears roamed. Three years later, Father Serra's expedition to the area was plagued by food shortages. Possibly because he knew of the Valley of the Bears (La Canada de Los Osos), Father Serra established two mission sites nearby: Mission San Luis

Obispo de Tolosa in 1772 and Mission San Antonio de Padua in 1771. Mission San Miguel was founded in 1797 after Father Serra's death. The two missions now located in San Luis Obispo County, Mission San Luis Obispo and Mission San Miguel, became known for their prolific vineyards and large landholdings.

Mission San Luis Obispo de Tolosa

Father Serra established the Mission San Luis Obispo de Tolosa on September 1, 1772, placing a holy cross at the site. According to historian Myron Angel, buildings with thatched roofs were erected on the site in 1773. The buildings were destroyed by a fire in 1776. Subsequently, all of the mission's new structures were roofed with red tiles made from a fired ceramic clay to protect them from fire. Today, you will find tile roofs on homes and

A drawing of the Mission San Luis Obispo de Tolosa by historian Myron Angel, circa the 1880s. *Original artwork by Myron Angel.*

buildings throughout California. In 1794, the mission was enlarged, and the quadrangle of buildings was completed in 1819.

Historian Myron Angel stated that the Mission San Luis Obispo de Tolosa soon became one of the most flourishing missions in California. Padre Luis Antonio Martinez is the person credited with organizing and managing the mission for thirty-four years. Padre Martinez was described as a man of comprehensive purpose and indomitable force by Angel. According to a survey of the mission dated 1804, there were 832 Native Americans living at the mission, and its livestock included cattle, sheep, horses and poultry. Gardens, olive trees and vineyards were also planted at the mission.

The life of the padre was not always harsh. Reverend Walter Colton, the alcalde of Monterey, wrote this description of Padre Martinez in his book, *Three Years in California* (1849): "His table was loaded with the choicest game and the richest wines; his apartments for guests might have served the hospitable intentions of a prince. He had 87,000 head of grown cattle, 2,000 tame horses, 3,500 mares, 3,700 mules, eight sheep farms averaging 9,000 sheep to each farm." When Padre Martinez returned to Spain, he took $100,000 with him as the fruits of his mission enterprise, according to Angel.

Winemaking at the Missions

The vineyards at Mission San Luis Obispo were planted prior to 1804. The Mission grapevine cuttings from Lower California were carried by the padres from mission to mission. The tradition of planting the vines is another legacy passed down from the padres to California wine growers. According to a written account dating back to 1858, the vines were spaced around five feet apart and kept closely trimmed, a method known as head pruning that was favored by the Spanish. This method shaped the vine to form a short, stocky trunk around four feet in height with a bushy low-growing vine. The thick trunk supported the vine, which could bear large amounts of fruit, with some bunches weighing more than three pounds. Most vineyards were protected by low adobe walls, rows of cactuses or close-set willow saplings.

The padres at the missions had both personal experience and access to early writings on viticulture and winemaking. The grapes were crushed immediately after the harvest in adjacent vineyards. Crushing the grapes was done by the local Native Americans who trampled the grapes, sometimes on cowhides. The juice was collected in ceramic vessels or leather bags and then emptied into stone vats and wooden tubs. There is evidence of a stone vat for fermentation and storage at Mission San Miguel. Supplies were often shipped to the missions in barrels, which provided the padres with onsite containers for storing their wine. The padres allowed the wine to ferment for two to three months before drinking.

By 1823, all but four Alta California missions were growing grapes and had flourishing wineries. The famous San Gabriel Mission was the largest, with three wine presses and eight brandy stills producing nine thousand gallons of wine and three thousand gallons of brandy annually at the height of its production. The Mission wine was used for sacraments in the Catholic Church and was served as table wine with meals. It was sometimes blended with the grape variety Alicante Bouschet, which was also grown in the mission vineyards, to deepen the color.

California's First Wine: Angelica

The first true California wine, Angelica, was made by the Franciscan padres once they obtained the necessary equipment—stills. Angelica was simple to make and was very sweet, as were most wines in the nineteenth century. No one knows when this wine was officially named Angelica, but legend has it that the name pays homage to the city of Los Angeles. Frenchman Emile Vachel traveled from France to visit Los Angeles; we have his notes, dated 1891 and written in French, on how to make Angelica wine. According to Vachel:

> *The basic technique was to mix the juice of freshly crushed Mission grapes with brandy, which was also made from Mission grapes. The recipe stated that the Mission grape is the only grape that should be used for Angelica. The proportions were three gallons of juice to one gallon of 180-proof brandy. The winemaking method stresses that must (freshly crushed grape juice containing seeds, skins and stems) should not be allowed to ferment. The must can rest in an open tank for 12 to 20 hours at a moderate temperature before mixing it with the brandy.*

In 1979, an 1875 bottle of Angelica was found, opened and judged as a superb old dessert wine by Harvey Steiman, the editor of *Wine Spectator*. He described it as the "most magnificent" California sweet wine that he had ever tasted—a wine that was 104 years old. Angelica is still made in small quantities by California winemakers; it appeals to those who enjoy sweet wines and early California wine history.

Mission San Miguel Arcángel

The Mission San Miguel Arcángel, the sixteenth of the twenty-one California missions, was established on July 25, 1797, by Franciscan Father Fermín de Francisco Lasuénas. The mission's location is thirty-four miles north of San Luis Obispo on the west bank of the Salinas River, near its junction with the Estrella River. A temporary church was built in 1797, but it burned down in 1806. The permanent church was built from adobe bricks composed of molded and baked clay soil.

A drawing of the Mission San Miguel Arcángel by historian Myron Angel, circa the 1880s. *Original artwork by Myron Angel.*

Father Juan Martin served this mission from 1797 to 1824; under his leadership, two prosperous vineyards were established in the area known as Vineyard Canyon. The first vineyard was planted between the mission and the Salinas River. The second vineyard was located at Vineyard Springs on eighteen acres northeast of the mission. In 1815, an adobe building consisting of three rooms was built in the middle of the vineyard to house the padre who was responsible for maintaining the vineyards. Winemaking took place at the mission, and the wine fermented in stone vats.

The End of the Mission Era and Spanish Rule

The Mexican War of Independence was lengthy, with many battles being fought after September 15, 1810, when Father Hidalgo, a Mexican priest, started the insurrection against the government of Spain. He fought with a small band of men; the new Mexican Empire was proclaimed on May 18, 1822. Alta California continued with a mission-dominated Catholic culture as it transitioned to become a Mexican territory.

In 1824 and 1826, the Mexican government passed laws to secularize the missions. Mexican government officials appointed administrators of these huge mission properties to distribute land to influential families and their friends. Dramatic changes took place as the orders for the secularization of the missions took effect between 1834 and 1836. Anger, frustration and violence followed. Mission buildings, crops, cattle and sheep were plundered by local Californians. Vineyards and orchards were often abandoned. The Catholic padres and priests realized they could not fight the loss of their land and churches. Many of the Catholic clergies returned to Spain, some taking riches with them.

However, the story of the Mission San Luis Obispo was different from those in many of the other mission towns. The mission was located in the heart of the community. Some mission buildings were sold to private owners and repurposed as barracks for soldiers, courthouses, local jails, schoolhouses, restaurants, saloons, lodgings and private residences. Orchards, vineyards and the surrounding lands were claimed by local people both prior to and after the occupation of the Americans in July 1846. Sixteen years later, the mission vineyards were purchased by Pierre Hypolite Dallidet, the first commercial winemaker in the county.

In the central and south areas of San Luis Obispo County, the outlying vineyards of Mission San Luis Obispo were abandoned. In North County, history tells a different story. The vineyards in San Miguel and Vineyard Canyon were maintained sporadically. The mission church was reestablished on the San Miguel property and remains a place of worship today. The wine continued to be used for the rituals and ceremonies of the Catholic Church.

Mission lands throughout the county became available for private ownership under the Mexican administrators, and ranchos came to dominate San Luis Obispo County. The agricultural heritage of Mission San Miguel Arcángel shaped the development of the North County; the early rancho owners continued to raise cattle and horses. Grain became a dominant crop. The era of the Mexican vaqueros established the art of horsemanship and roping to maintain herds. Today, that heritage is celebrated throughout San Luis Obispo County at the local fairs and rodeos. Large vineyards were not to be reestablished until the 1880s.

The Mexican-American War: California Becomes a State

During the presidential campaign of 1844, James Polk promised the westward expansion of United States territory through the annexation of Texas. After his election, Congress claimed the area north of the Rio Grande, the border of Mexico at the time. The United States was interested in acquiring the resources available in California.

The Treaty of Guadalupe, signed in 1848, ceded California, New Mexico and all the land west of the Rocky Mountains to the United States. However, life in California continued as before—that is, until news began to spread that gold had been discovered at Sutter's Mill near Sacramento, California, in the northern part of the state. The discovery of gold in 1848 changed the world and accelerated the spread of viticulture in California; over three hundred thousand prospectors came looking for their fortunes and some jug wines to drink.

California became a state in 1850; San Luis Obispo was designated as one of the state's original twenty-seven counties.

Historic Sites to Visit

Mission San Luis Obispo de Tolosa
Mission San Miguel
Point Luis Lighthouse
Morro Rock
Piedras Blancas Lighthouse
Rodeos at the California Mid-State Fairgrounds

2

From Seeking Gold to Growing Grapes

The Legends of San Luis Obispo County

GRAPE VARIETY: ZINFANDEL

Introduction

A lump of gold the size of a potato was displayed in New York on the steps of the stock exchange to show proof of the discovery; the *Baltimore Sun*'s first reports of finding gold in California were published on September 20, 1848. There is no doubt that the discovery of gold in California profoundly shaped the history of nineteenth-century America. This astonishing news made San Francisco the destination of travelers around the world. Men and women planned to emigrate to California to seek their fortunes. They came by ship around Cape Horn of South America and by land across the Isthmus of Panama and through Mexico to reach ports on the Pacific Coast, including San Francisco Bay. They also came by foot and wagon from the Midwest, across the plains, over the Rocky Mountains and through New Mexico and Sonora to California.

Nurserymen, horticulturists and viticulturists followed these fortune seekers with plans to make their own fortunes by importing nursery stock for gardens, orchards and vineyards. The thousands of gold seekers needed food and drink. These men, including James L.L. Warren, Anthony P. Smith and Captain Frederick W. Macondray, traveled from New England to explore the geography of California. They mapped out the areas around San Francisco Bay to study the favorable soils and climates.

These merchants and nurserymen, importers and growers of deciduous fruit trees and grapevines, soon established commercial nurseries to encourage the development of agriculture between the towns of Sacramento and Fresno. They published agricultural journals and formed local horticultural associations to recruit growers.

Over three hundred thousand emigrants settled in California in the five years following the discovery of gold in the state. Most did not find gold, but many found jobs in agriculture, food and wine production. Some were able to purchase their own land, where they farmed and planted their own vineyards.

The Impact of the Gold Rush on California

At the beginning of 1848, the California Territory was sparsely populated. The Native American populations had been reduced to approximately 15,000 during the Spanish and Mexican occupations. There were around 6,500 Californios of Spanish or Mexican heritage and around 700 foreigners from the American territories and states living in the California Territory at the time.

Then, gold was discovered on the American River near Sacramento by James W. Marshall, a carpenter who was building a water-powered sawmill for his boss, John Sutter. The two men made a pact to become partners and keep their discovery a secret. This discovery of gold occurred on January 24, 1848, and six days later, Mexico ceded California to the United States per the Treaty of Guadalupe.

As news of the discovery spread across the world, thousands of people seeking their fortunes arrived in Northern California. By August 1848, four thousand gold miners were in the goldfields, and they were soon followed by eighty thousand "forty-niners," the fortune seekers of 1849. California became a state in 1850; San Luis Obispo was named one of the state's original twenty-seven counties in the same year.

By 1853, immigrants were experiencing primitive living conditions, backbreaking work and extraordinarily high costs of living. The amount of gold extracted in the state was valued at over $2 billion, but very few prospectors "struck it rich." As foreigners arrived, Californians developed prejudices against them—particularly the French—and drove them from the goldfields. Laws were passed to restrict gold claims owned by "foreigners." The California gold rush peaked in 1852 and had dissipated by the end of the decade. Many immigrants were left stranded, seeking new possibilities.

San Luis Obispo County Settled by Farmers from Around the World

As gold lost its luster, farmers and dairymen moved south to San Luis Obispo County. The county was unique in that many of the immigrants who settled there in the 1800s were able to homestead or purchase their own land to farm grains, fruits and nuts. Some chose to purchase ranchos to raise cattle and sheep. Others settled along the coast and established dairies or built fishing boats. Everyone had chickens, gardens and a few grapevines to provide food and wine for their own tables. Australian, Danish, Dutch, English, French, German, Italian, Mexican, Swedish and Swiss people and Mennonites all settled in close proximity, sharing their cultures, labors and friendships. Descendants of those early pioneers still live in the area; some still farm the land they inherited or purchased from their families.

Very few settlers had worked in viticulture or commercial winemaking before moving to California, but wine was part of their food culture and religion. French, German and Italian families made their own table wines at home. They planted a few rows of vines in their gardens, primarily Zinfandel, and enjoyed their own wine daily.

Within two decades, legendary pioneers from all over the world, the first commercial growers and winemakers in San Luis Obispo County, were making wine history.

Pierre Hypolite Dallidet: Grower and First Licensed Winemaker and Distiller (1822–1909)

Pierre Dallidet is celebrated as the first commercial grower and winemaker in San Luis Obispo County. He was the first licensed commercial distiller in the county as of 1891. His legend is one of adventure, political upheaval, military service, land development, viticulture, wine production and a tragic murder within his family.

Dallidet was born into poverty in 1823 in southwestern France. At the age of twenty-three, his travels took him around the world, and he served in the French military in Tahiti, located within the Society Islands, a part of French Polynesia. Dallidet enjoyed the peaceful environment of the island and its barter economy; he trained as a carpenter in the military and saved his salary for future investments abroad.

When he completed his military service on December 31, 1850, Dallidet sailed with other soldiers to San Francisco to stake his own gold mining claim in a place called Hangtown (now known as Placerville). But within a few years, he was walking south toward Mexico with a group of his countrymen to escape anti-immigration sentiment in Northern California. California was transitioning to an Anglo-American society with little tolerance for other ethnic groups, especially those seeking their fortunes in gold. The Mexicans, Chinese and French were all persecuted with brutal physical punishments and restrictive laws that prevented them from owning land or working mining claims. The "Foreign Miners License Tax" required non-citizens to pay twenty dollars per month to work their own claims.

Dallidet started his perilous journey from Hangtown to San Luis Obispo with approximately 150 French "soldiers of fortune" who decided to move south to Mexico in 1853. As they passed through the Central Coast, some decided to settle in San Luis Obispo, and others chose to stay in Santa Barbara. Dallidet recognized that the land was fertile and that the area had a Mediterranean climate similar to France. He saw the opportunity to make a new life for himself, first as a carpenter, then as a landowner and farmer. He dreamed of making his fortune by exporting his produce to the outside world.

In 1859, Dallidet married Marie Ascension Salazar and purchased approximately sixteen acres of land adjacent to the town's old withered mission vineyard. He built his own adobe home over a unique wine cellar that was designed in the French style. Dallidet planted orchards and new vineyards before building his winery near his home.

His was the first winery to be licensed in the county. By 1883, his seven-acre vineyard contained 7,200 vines, producing 3,300 gallons of wine per year. Dallidet made $250 per acre per year from grape and wine sales in 1884, according to the town's newspaper, the *Telegram-Tribune*. In 1886, he sold 635 gallons of Chardonnay and 2,717 gallons of Mission wine. By 1889, his vineyard had doubled in size, with eleven acres planted with wine

Pierre Hypolite Dallidet, the first commercial winemaker (he worked from 1860 to 1900) and first licensed distiller in 1891 in San Luis Obispo County. *Courtesy of the History Center of San Luis Obispo County.*

grapes and three with table grapes. From the 1860s to the 1890s, Dallidet was well known for his pioneering commercial wine and brandy making, sourcing his fruit from his own vineyards and orchards. Dallidet also served in the local government; his home, a center for the culture, attracted artists, archaeologists, writers and musicians from all over the world. His children were educated, artistic and talented.

He became famous for planting over two hundred varieties of grapes, sharing his research and rootstock with others. Dallidet collaborated with local vintners, including Henry Ditmas and A.B. Hasbrouck of the upper Arroyo Grande Valley. From one of his son's diaries, we know that Dallidet planted Charbono, Black Malaga, Muscat of Alexandria among many other grape varieties. He sold cuttings to other vineyardists and consulted with them on grafting cuttings with the Mission grape rootstock, which appeared to be hardy and resistant to disease. The vineyards in France were decimated by the louse phylloxera beginning in the 1870s, and Dallidet assisted the French government by grafting cuttings of French varieties onto his own disease-resistant rootstock.

Dallidet bottled his wine and sold it locally as well as throughout the state of California. He was the first winemaker in the county to make blended wines that were known for their quality. The *Daily Republic*, in an article published on Thursday, October 3, 1889, reported:

> *Until the last two or three years, there was scarcely any wine made except by Mr. Dallidet, who usually made about 6,000 or 8,000 gallons annually and a few hundred gallons of brandy. Wine made from the Mission grape and bottled twenty years ago by Mr. Dallidet is now equal to the best Chateau wine of France. This is proof that age and skillful bottling are the chief factors in making good wine in this locality.*

A map of vineyard and orchard plantings from the diary of Louis Pasqual Dallidet, the son of Pierre Hypolite, dated 1882. *Courtesy of the History Center of San Luis Obispo County.*

Dallidet kept financial records and noted that during the financial boom of the 1880s, his wine made a profit of twenty-five cents per bottle, but the cost of shipping was nine cents. His brandy, which was shipped in the same size bottle, yielded a seventy-five-cent profit per bottle with the same cost of shipping. Dallidet applied for and became the first licensed distillery in San Luis Obispo County in 1891—the only distillery in the county prior to World War I.

Dallidet developed real estate projects with his brother-in-law Victorino Chaves. He acquired 150 acres suitable for viticulture and farming, including all the mission-era vineyards in what is now known as downtown San Luis Obispo. By the mid-1880s, Pierre Dallidet was the largest single landholder in the city.

Pierre Hypolite Dallidet died in 1909. His later life was marred by financial ruin, family disputes and the tragedy of one of his sons murdering another. His last surviving son, Paul, willed the family home and garden, which had been occupied for over one hundred years by two generations of the Dallidet family, to the City of San Luis Obispo in the event of his death in 1958. It was recognized as California Historical Landmark No. 720 in 1960.

James Anderson: Shipwreck Survivor and Wine Grower (1852–1921)

James Robert Anderson, a shipwreck survivor from Sydney, Australia, was one of the founding viticultural pioneers of San Luis Obispo County. He was the first Australian to settle on York Mountain, the first to grow Zinfandel and Burger grape varieties, the first to build a winery in North County and the first to own a redwood tank with a sixteen-thousand-gallon capacity. He pursued a formal education in agriculture, studying climate and soils, which was unusual for the times.

The lure of California gold brought thousands to the state's shores. Many were fleeing from poverty, oppression, drought and war. The Anderson family was an exception to this, as they had made their fortune in sheep raising in Australia. The patriarch, Andrew Anderson, and his wife decided to bring their fortune of over $100,000 to California to pursue new dreams.

The family's dangerous journey began in 1855, when Andrew Anderson bought passage to San Francisco for his wife, Elizabeth, and eight children on the ship *Julia Ann*. The ship was wrecked in the Society Islands. Many lives were lost, including one of the Andersons' children. The family's possessions, including the $100,000 fortune, sank to the bottom of the sea. The ship had hit a reef, ran aground and did not sink but was broken up on the rocks. Eventually, the family was rescued and sailed on to San Francisco. The Andrew Anderson family started over, settling on a farm in San Jose. The children went to school in Santa Clara County; one of the Andersons' sons, James, was formally educated in horticulture and agriculture. He traveled all over the Pacific Coast to study soils and climate. James farmed a variety of crops in several counties and states before he settled on the Central Coast in 1876 to farm grain. He continued his studies of viticulture and winemaking and decided to plant his own vineyard.

James Anderson bought Mr. Dunn's farm, comprising 163.75 acres in the Ascension District at the base of what is now known as York Mountain, in 1879. Anderson's land was located on Anderson Creek (which was later named after him), about halfway between Templeton and the Pacific Ocean. Anderson set about clearing his land, planting an orchard and vineyard on 20 acres.

In late 1879 or 1880, Anderson became the first to plant a white wine grape variety, Burger, in the Ascension District. He also planted Zinfandel, the most commonly planted red wine grape in San Luis Obispo County at the time. In 1882, James Anderson built the first winery on York Mountain and carefully selected the finest equipment for his venture. He envisioned a large production facility and purchased a sixteen-thousand-gallon redwood storage tank. The tank was made in San Francisco, purchased in the San Jose area and received much publicity on delivery.

James married Miss Lizzie Gray in Bakersfield, California. They had six healthy children, five of whom eventually settled near their parents. Their son Frank joined his father in the vineyards and the winery business, producing wines until the onset of Prohibition in 1920.

James Anderson became a prosperous and well-respected viticulturist in San Luis Obispo County. The 1891 Directory of Grape Growers, Wine Makers and Distillers of California lists him as a winemaker. The directory states he farmed twelve acres of bearing wine grapes and recorded two varieties: Zinfandel and Burger.

James Anderson's wife died in 1899; James lived until 1921 and was buried in the Cayucos Cemetery. The Andersons' son Frank, who attended the Pacific Coast Business College, worked with his father, managing the ranch until James's death.

THE LEGEND OF HENRY DITMAS: SAUCELITO CANYON, WILLOWS AND ZINFANDEL (1845–1892)

Henry Ditmas was the first to plant and dry-farm Zinfandel grapes on a three-acre block in the Upper Arroyo Grande Valley. His vineyard, nestled in a canyon at 860 feet above sea level, dated back to 1880. His Zinfandel grapes were first sold as table grapes, winning awards in local fairs. Ditmas sold wine grapes to his neighbor A.B. Hasbrouck, a winemaker at St. Remy Winery, the first production facility in the South County.

Englishman Henry Ditmas planted the first Zinfandel vineyard in Arroyo Grande Valley in 1880. He named his land Rancho Saucelito after the willow trees. The Spanish word for willow is *sauce*. *Courtesy of the Greenough family.*

Henry was the son of Englishman colonel Thomas Ditmas, who was stationed as governor of a British military cantonment at Secunderabad in British East India. Henry was born there in 1845 but was sent to England for his education. He graduated from college as a civil engineer at the age of eighteen in 1863.

Henry's first job was as an assistant engineer to a British army officer building military roads in New South Wales, Australia. Henry was selected to supervise and complete the project when his boss became ill. Henry was highly regarded for his accomplishments in completing the project. In 1866, he received a "letter of recommendation as an engineer" from the British Colonial Office, which launched his career in the military.

At the age of twenty-one, Henry was sent to British East Africa to build a military road through the jungle. He supervised two hundred Black Africans and a Portuguese crew. Although he initially shared no common language with his workers, Henry mastered Portuguese. He also learned a number of African dialects so he could teach the African workers the skills they needed for the challenging job.

Henry contracted malaria in British East Africa, forcing him to leave the military and return to England for medical care. While in England, he spent time painting watercolors of English sailing ships and met the woman he would eventually marry, Rosa. Henry gradually recovered from malaria but began to suffer from rheumatism. Like many others, Henry decided to immigrate to the United States to seek his fortune and a warmer climate.

In 1871, Henry and his cousin Eben traveled to San Francisco by ship to pursue the sheep and wool trade. On docking, they met a man named Benriamo, a hotel owner in Avila Bay who convinced them to settle in San Luis Obispo County, which Benriamo claimed was the finest sheep country in the world. The cousins worked for a few months for various sheepherders in the Arroyo Grande area to learn the business. They leased pastureland in the Los Osos Valley and invested in a few thousand sheep.

Henry filed citizenship papers before returning to England to marry Rosa in 1874. The couple took a long honeymoon cruise to California via the Isthmus of Panama. They arrived in the Avila Bay at Port Harford. Their son, Cecil, was born on November 7, 1875.

The year 1877 brought severe drought to California. Henry sent his sheep to pasture in the Sierra Nevada Mountains, but most of the flock died in a spring snowstorm. In 1878, Henry filed a government claim for 560 acres adjoining Ranchita Arroyo Grande, which was owned by the famous cattlemen the Steele brothers. He named the land Rancho Saucelito, a Spanish name in honor of its many bordering willow trees. Henry built a wood-frame cottage on the site, and Rosa planted an English flower garden. The following year, 1879, Henry cleared the land; Zinfandel and Muscat grapevines were planted in 1880. The vines were sent from Europe and purchased locally, according to the Ditmas family. This is the first documented planting of Zinfandel in the Upper Arroyo Grande Valley. Horticulturist David F. Newsome later gave Henry additional grapevines that were suitable to the warm climate of Rancho Saucelito. Rosa Ditmas pursued cattle ranching; the Ditmas livestock brand, the broad arrow, was recorded on April 23, 1883, at the request of Rosa C. Ditmas for her exclusive use in branding her stock.

The first harvest fair was held in Arroyo Grande in 1885. There, Henry sold table grapes and raisins to his neighbors and other locals in Avila Bay and Arroyo Grande. Both the Upper and Lower Arroyo Grande Valleys produced the finest vegetables and fruits in the county.

Rosa and Henry's marriage failed in 1886; they chose to divorce. Rosa continued to live in their cottage in Upper Arroyo Grande Valley, where she managed the cattle ranch and vineyards and cared for Cecil. Henry moved to San Francisco after deeding his property to Rosa. In 1892, he died from complications of malaria and pneumonia in Boston, Massachusetts. At the time, Henry was only forty-seven years old and was fondly remembered as an artist, a photographer, a world traveler, a civil engineer, a military officer, a businessman and an extraordinary wine grower.

Abram Bruyn Hasbrouck: St. Remy Winery (1845–1915)

The first winery in the Upper Arroyo Grande Valley, St. Remy, was built by a man from a wealthy family of Dutch Huguenot ancestry who settled in upstate New York. He dreamed of sailing the seas and exploring the Wild West. He had a style and grace that made him a legend in hospitality on the Central Coast. His beautiful English wife, an accomplished pianist, owned her own cattle brand and was famous for her love of roses and English gardens. He designed and built a home with a teahouse and elegant gardens that attracted visitors from far and wide. His name was Abram Bruyn Hasbrouck, but he was known as A.B.

A.B. was greatly influenced as a child by the seafaring adventures described in the book *Two Years Before the Mast* by Richard Henry Dana. A.B. went to sea as a teenager and developed a passion for mapping the

St. Remy, the home of A.B. Hasbrouck, was considered the grandest house built on the Central Coast. Built in 1884, the property was planted with lawns, trees, hedges, roses and flower gardens. Hasbrouck was famous for his hospitality. *Courtesy of the Greenough family.*

world. He pursued his dreams of exploring the American frontier after the Civil War and the death of his father. He traveled to Colorado and quickly lost over $10,000 in a mining investment. He then journeyed to San Luis Obispo County in 1866, seeking employment as a vaquero with the famous cattlemen George and Edward Steele. The Steeles owned two ranchos, Corral de Piedra (22,000 acres) and Bolsa de Chemisal (13,600 acres), in the Arroyo Grande Valley. They offered him a job in their dairy operation. A.B. had the entrepreneurial skills he needed to become the majordomo for the Steeles' businesses before striking out on his own.

A.B. started his own dairy farming business, Ranchita Arroyo Grande Cheese, in 1873 on the 4,437-acre Rancho Arroyo Grande property, which he leased from the Steele brothers for the next ten years. He also organized the first cheese cooperative in the area with other dairymen. A.B. eventually bought Rancho Arroyo Grande from the Steele brothers for $27,000 in 1883 and established his St. Remy Ranch. The property was adjacent to that of grape growers Henry and Rosa Ditmas. Henry, Rosa and A.B. became friends and shared their expertise with one another.

A.B. built a grand home and planted lawns, hedges and flower gardens. He built a summer house and furnished it with pieces made from the native woods growing on the rancho. A.B. invited his friends for picnics and luncheons at the ranch; although it took four hours by carriage from the coast to arrive at his home in the Upper Arroyo Grande Valley, hundreds of visitors joined him throughout the years.

In 1884, he started work on his St. Remy Winery, building the stone foundation. A.B. planted his thirty-acre vineyard with a number of varieties after consulting with other viticulturalists in the county. He hired Chinese and Portuguese workers for his new enterprise and was famous for feeding them well. He was the first person in the county to hire Chinese workers to work in his vineyards.

The following year, 1885, Henry Ditmas sold his grapes to A.B. Hasbrouck, who began making wine at St. Remy Winery under his own label, St. Remy. His wines won first place in the 1896 California State Fair. Ditmas and Hasbrouck understood that a hardy rootstock was necessary for a grapevine to survive in California. Henry Ditmas's granddaughter confirmed that both men bought rootstock from Pierre Hypolite Dallidet and grafted cuttings of Zinfandel and Muscat onto this rootstock. Both men studied and followed the writings of Sonoma viticulturist Agoston Haraszthy, who advocated for planting these varieties. In 1887, the county board of trade disseminated a pamphlet that contained an article written by P.H. Dallidet on the conditions

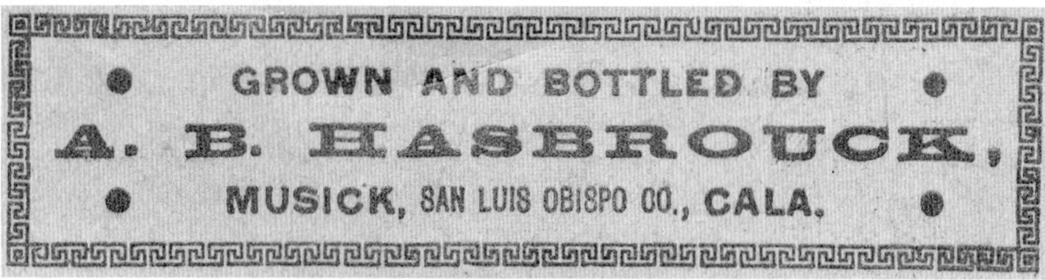

Top: St. Remy was the first winery built in the Arroyo Grande Valley. Its owner, A.B. Hasbrouck, was famous for serving fine food to his Chinese and Portuguese vineyard workers, who are shown here in front of the winery, which was built in 1884. *Courtesy of the Greenough family*.

Bottom: The A.B. Hasbrouck wine label was first produced in 1886 at St. Remy Winery from the Zinfandel grapes he purchased from his neighbor Henry Ditmas in the Upper Arroyo Grande Valley. *Courtesy of the Greenough family*.

Opposite: A.B. Hasbrouck and his wife, Rosa Ditmas, at their home in the Arroyo Grande Valley, circa late 1880s. *Courtesy of the Greenough family*.

of the wine industry; Mr. Dallidet's article stated, "I am of the opinion that the wealth of San Luis Obispo County can and will be greatly increased by the planting of vineyards. Mr. Hasbrouck of the Ranchita and Mr. Ditmas of Musick each have vineyards."

Rosa married A.B. a year after her divorce from Henry and moved into the St. Remy house. They raised Cecil Ditmas as their son. A.B. was well

known for his cheeses, wines and extraordinary vegetable gardens. He became a local celebrity. His visits to Avila with gifts from his ranchita were recorded in the local newspapers. He ultimately suffered financial ruin, losing his ranchita to mortgage holders. A caring friend purchased the ranchita acreage, including the winery, gardens, vineyards and A.B.'s beautiful home, and gifted it to A.B. to enjoy for the rest of his life.

The vineyards on A.B. Hasbrouck's ranch continued to flourish until 1915, when phylloxera destroyed them. A.B. died in the same year. Rosa leased the Saucelito Vineyards to various tenants until she died in 1927.

A local sheriff harvested grapes and produced wine during Prohibition. The wines made from the grapes planted by the legendary Henry Ditmas were noted for their quality and superior flavor. Cecil inherited both properties from his mother at her death and became a cattle rancher.

Historic Sites to Visit

Dallidet Adobe and Gardens in San Luis Obispo
Cayucos Morro Bay Cemetery
Old Mission Catholic Cemetery in San Luis Obispo
South County Historical Society
Avila Beach and the Bob Jones Trail
Dana Adobe
Village of Arroyo Grande
Lake Lopez
Pozo Saloon

3

"Go West"

Trains, Harbors and New Prospects in Agriculture

GRAPE VARIETIES: ZINFANDEL, BURGER, MUSCAT OF ALEXANDRIA, BLACK PRINCE, BLACK HAMBURG, BLACK MALVOISE, GOLDEN CHASSELAS, WHITE TOKAY, CHASSELAS DE FONTAINEBLEAU, FISHER ZAGOS, BLACK MALVASIA, BLACK MOROCCO, ROSE OF PERU, FLAME TOKAY AND EARLY VICTORIA

Introduction: Trains and Harbors Bring New Pioneers, Leading to Decades of Population Growth, Viticulture and Education

The building of the world's first transcontinental railroad, known as the Pacific Railroad, between 1863 and 1869 provided the transportation needed to bring new settlers to San Luis Obispo County as the nation expanded westward. The opening of the railroads changed American society and commerce. The U.S. economy expanded as traveling and shipping goods became cheaper. At this time, settlers also saw new possibilities in agriculture. Cattle, grains, fruit, grapes and wine were shipped throughout California as soon as railroads were routed into the state.

While the railroad routes were under construction, goods and passengers traveled by ships and steamers along the Pacific Coast, docking in San Luis

Obispo County at San Simeon, Cambria, Cayucos, Morro, Spooners Cove and Port Harford in Avila Bay. Most of the new arrivals wanted to buy land in the county's warm climate and build wealth through agriculture and commerce. Many had no experience, but they all found educational resources to help them succeed.

The State of California formally organized those educational resources for farmers and vineyardists. James Warren, the "Father of California Agriculture," published newspaper articles and agricultural journals; he also organized local, regional and county fairs for agricultural competitions. In 1854, the California Legislature voted to establish an annual exhibit of the state's flowers, fruits, grains, livestock and vegetables. The first state fair was held in San Francisco on October 4, 1854. It has since become an annual two-week event (the event was only missed during World War II and the 2020 pandemic). The California Fairground in the state capital is the place where major grape and wine competitions have honored quality and held both growers and winemakers to the highest standards for 150 years.

The Morrill Act of 1862, which was signed by Abraham Lincoln, funded educational institutions by granting federally controlled land to establish and endow "land-grant" colleges, which would focus on teaching practical agriculture, among other subjects. Governor Henry H. Haight signed the Organic Act to create the University of California (UC) on March 23, 1868; the UC campus at Davis emerged after Prohibition as the leading research and training institution for enology and viticulture in the nation. The Hatch Act of 1887 provided federal funds for states to establish a series of agricultural experiment stations. UC planted experimental vineyards and new grape varieties and studied pest control management in San Luis Obispo County. Since 1889, this research has been shared with farmers and vineyardists by the county agriculture advisor, who continues to offer these services today.

Viticulture became an economic force in California, expanding rapidly across the state with each new wave of settlers. Grapes were a successful crop from the mountains of the High Sierra to the deserts east of Los Angeles. San Luis Obispo County is located halfway between two of the most important historic wine-growing areas: the San Francisco Bay area, including Napa and Sonoma Counties, and Los Angeles. A wide variety of grapevines, which were imported from Europe in the nineteenth and twentieth centuries, are successfully grown throughout the county. For the first one hundred years, most of the county's grapes were sold to outside areas. When the California wine revolution arrived in the county in 1972, these premium grapes began producing award-winning local wines. However, the road to recognition and fame was difficult to navigate; few knew anything about San Luis Obispo County, but four wine legends changed everything.

Grape Growing Is Firmly Established in the 1870s

According to an article published in the *Democratic Standard* on October 15, 1870, titled "Our County—Its Climate and Resources," there were four grape growers in San Luis Obispo County. Mr. D.F. Newsome grew Muscat of Alexandria in Arroyo Grande; J.P. Andrews owned a vineyard planted with "Old Mission varieties" alongside "several of the choicest foreign kinds"; W.T. Sheid in the Estrella Valley grew Black Prince (later identified as Zinfandel) and Muscat of Alexandria, among others not specifically named; and the Messrs. Dore were "resuscitating and improving" an old Mission vineyard at the Rancho of San Ysabel.

"The Profits of Grape Growing" published in the *Democratic Standard* on January 28, 1871, noted that Mr. Shaw was "making a good profit on his Muscat of Alexandria vines." The Meister brothers were reported to "have a great variety of grapes that led to high profits," including Muscat of Alexandria, Black Hamburg, Black Malvoise, Golden Chasselas and White Tokay. The article described the Mission grape as abundant but the "least valuable."

On June 17, 1876, the *Tribune* reported that George W. Hampton was noted to have a vineyard of a "thousand vines" covering two acres of his farmland. He planted nine varieties of grapes but no Mission grapes. His foreign varieties produced a higher yield and better profit: Muscat of Alexandria, Chasselas de Fontainebleau, Fisher Zagos, Black Malvasia, Black Morocco, Rose of Peru, Flame Tokay, Early Victoria and Black Hamburg.

The California Board of State Viticultural Commissioners was established in March 1880 to promote viticultural industries throughout the state. Most new growers were uncertain about what varieties to plant in their new vineyards. The board provided education for planting, pruning, fertilizing, fermenting, distilling and treating diseases of the vine. Newspapers and journals also published articles on viticulture.

Wine Legends Arrive from the Midwest

Many settlers from the Midwest came to San Luis Obispo County in search of new land to farm during this period of wine history. Wine legend Andrew York from Indiana settled in the Santa Lucia Mountains in North County, west of Paso Robles. East of Paso Robles, twin brothers John and William Ernst settled in the hills of the Geneseo District, near Creston, and they were followed by Gerd Klintworth, a German immigrant from the famous Anaheim colony of wine growers in Los Angeles County. These four men studied and experimented with grape varieties before establishing their wineries.

The potential of viticulture in San Luis Obispo County during the 1880s was described in the following terms: "The warm area east of the Santa Lucia is most favorable to the growing of the grape, there it being the richest and ripening the earliest, but in sheltered and favored localities near the coast, the grape grows as well as in any other portion of California, although ripening later." Many advertisements were placed in small-town newspapers across the Midwest to entice people to "go West" and settle in San Luis Obispo County.

The county also became a place of interest for fruit growing. In the early days of the county's mission history, the padres planted orchards of olive, fig and citrus trees, which thrived until the 1830s. Fifty years later, land salesmen predicted that every locality could successfully grow all types of fruit, including citrus, seed and stone fruits, without irrigation. Farmers came to buy land with the hopes of selling these crops to markets beyond local boundaries. The county assessors' early records, dating back to 1873, reported that forty thousand acres were under cultivation in San Luis Obispo County. Fruit trees and vines were counted by the number in the early days rather than by the acre. Sixty thousand grapevines were planted in 1873, and by 1876, that number had increased to eighty thousand.

Self-Made Men Invest Their Fortunes in Viticulture

In the early 1880s, investors who arrived in South County, near the city of San Luis Obispo, to purchase land were educated and wealthy. These men, who had made their fortunes elsewhere, came to replicate their success in the growing table and wine grape industry. One of the most famous of these men was the Honorable Frank McCoppin, who was noted as the first foreign-born mayor of San Francisco. According to the April 30, 1881 edition of the *Tribune*, McCoppin had "purchased the property located near this city heretofore belonging to the Bank of California." The following year, the *Tribune* reported on March 25 that McCoppin was "preparing to set out the largest vineyard in this county" with "160,000 cuttings." The vineyard was located on the western slope of the Santa Lucia Mountains in Van Ness Canyon.

Dr. W.W. Hays, one of the first physicians in the county, planted a large vineyard just north of the city of San Luis Obispo. Dr. Hays's visit to the McCoppin vineyards was reported in the *Tribune* on May 22, 1883; in the article, the vineyards were described as "recently planted" vineyards that consisted of "over 70,000 wines."

The Board of State Viticultural Commissioners published the Directory of Grape Growers, Wine Makers and Distillers in 1888 and 1891. The first edition listed growers', makers' and distillers' names with the counties and towns where the vineyards were located. The 1891 edition confirms that McCoppin held forty acres of fruit-bearing wine grapevines in San Luis Obispo. He is listed as a wine grape grower, but not as a winemaker.

The Legacy of Andrew York (1833–1913)

Andrew Jackson York first traveled to California at the age of twenty-one with his brother, Eli McLane York, in 1854. They headed west in a wagon train with a team of oxen, helping to drive over seven hundred head of cattle and fifty horses and mules. Their lineage can be traced back to England, but they grew up with Midwest values and a love of farming.

His Legend

The brothers, "Mac" and Andrew, had a series of adventures; they worked in the mines of Nevada but found no riches, so they moved on to California to try farming in Napa. Mac settled permanently in Napa to farm and establish a nursery to supply plants to the growing population. He planted Zinfandel in his vineyard and built his own York Winery in 1880. An article in the *St. Helena Star* dated October 27, 1884, described his grape harvest:

> *E.M. York has an immense crop of his own grapes—230 to 300 tons—and was absent, getting more help to pick them, at the time of our visit. His young son informs us that some of his vineyard yielded as high as twenty tons per acre.*

Mac's vineyards flourished until the phylloxera epidemic destroyed the vines; Mac visited with Andrew often, offering advice on selecting vines, equipment and winemaking.

Andrew moved around the United States and experimented with a variety of farms and crops over the next twenty years. Tragedy struck when fire took the life of his wife, Louisa, at their Missouri farm in 1873. With five motherless children to raise, Andrew sold his farm and visited Mac in Napa before moving to San Luis Obispo. While growing grain in Los Osos Valley, Andrew met a young woman who was divorced and living with her parents and her children. Their friendship blossomed; Huldah Matthews married Andrew in San Luis Obispo on September 21, 1876. This decision stabilized both of their lives, leading to financial security for generations. Huldah had an extensive knowledge of plants and their medicinal qualities. She became well known for her home remedies and served as a midwife in the York Mountain area, delivering babies until she retired at the age of sixty.

As viticulture began to thrive in California in the 1880s, Andrew and Huldah York discovered a property of 112 acres for sale by the owner located in the Ascension School District, now known as York Mountain. The property included a residence, fruit stand, vegetable garden, fruit trees and a small vineyard planted with Mission grapes. The owner, blacksmith Jacob Grandstaff, homesteaded the land in 1875. The Yorks bought the property and moved to the site, which is now one of the most historic and important sites in the county's wine history. The York family occupied this land, grew grapes and made Zinfandel wine for three generations, from 1882 to 1970. It is the longest-running family-owned winery in the history of San Luis Obispo County.

Andrew cleared more land and added trees to the orchards. He planted and dry-farmed forty acres of vineyards in 1883 with a variety that was soon to be known as the "heritage grape of California" Zinfandel. According to his grandson Silas York, "By 1886, the vineyards were extensive, and the

The York family grew grapes and made Zinfandel wine for three generations, from 1882 to 1970. It is the longest-running family-owned winery in San Luis Obispo County. *Courtesy of Neil Abbey.*

cuttings came from Napa." Andrew York probably obtained some of his Zinfandel cuttings from his brother, Mac.

Andrew sold his grape crops directly to his customers in the 1880s, but as his vineyard yields improved, he accumulated a surplus of grapes. He studied the books and articles that were in print at the time to learn about winemaking methods. Most likely, Andrew read the popular manual *The*

Andrew and Huldah York purchased their farm from Jacob Grandstaff in 1882. Andrew ran the vineyards and the winery. Huldah was famous for her knowledge of medicinal plants; she was a midwife and delivered a number of babies on York Mountain. *Courtesy of Neil Abbey.*

Wine Press and The Cellar, which was written by E.H. Rixford in 1883. This book was still a part of the Yorks' personal library in 1993. Vicki Dauth, a local historian, remembered that Rixford offered all kinds of viticulture advice, including suggestions that "egg white be added to the wine for fining" and that "wines not be racked under a full moon or during a storm." By 1891, Andrew York was listed as a winemaker with a vineyard of forty acres growing Zinfandel and Burger varieties and producing sixty tons of grapes in Templeton, according to the Directory of Grape Growers, Wine Makers and Distillers of California.

Andrew engaged his sons James and Thomas and stepson, Justus Priest, by having them collect the boulders that were strewn around the property in order to start the construction of the wine cellar adjacent to a hill. By 1895, the two-story winery was completed; it was built over the stone cellar and designed so that the grapes could be unloaded and crushed on the second floor. Using gravity, the juice traveled to the first floor to be stored in redwood tanks for fermenting. The wine was then placed in barrels and sold to the public. According to Andrew's son Silas York:

York Brothers Winery is located at an elevation of 1,500 feet and is seven miles from the Pacific Ocean. It became the largest winery in the county and was producing one hundred thousand gallons of wine annually by 1911. *Courtesy of Neil Abbey and Jan York.*

> *Andrew purchased fifteen 1,000-gallon tanks at $16.00 each, a hand crusher and an old press for $75.00 from his brother, Mac York, the wine grower in St. Helena. Eli had no further use of the equipment because phylloxera had gotten into his vineyard and he had to pull up the vines and plant prunes.... The fifteen tanks took up more room than anticipated and so the barrels had to be kept in a dugout under the family residence nearby.*

Redwood barrels were commonly used in California. Tall redwood trees provided lumber to make tanks holding between five thousand and sixteen thousand gallons of wine. *Courtesy of Neil Abbey.*

The winery was first known as the Ascension Winery; the name was later changed to A. York and Sons. James, Thomas and stepson Justus were involved in the construction, harvesting and winemaking by 1896. Three years later, Thomas sold his interest to his brother Walter, who continued to work with his father, Andrew, until his death in 1913. In 1902, Silas York was admitted to the partnership.

Silas described York Winery customers, saying they were "mainly teamsters, who, having hauled their loads up one of the steep grades on either side of the [York] mountain, had to pause to refresh their horses at the water trough placed there for their convenience." The Swiss, Italian and Portuguese immigrants from Cayucos and Cambria were also loyal customers. Within a few years, the small winery was producing 1,500 gallons of Zinfandel wine annually. According to wine historian Charles Sullivan, Andrew York developed a good reputation for his table wine.

The winery broke all local records for production in the early twentieth century. A. York and Sons became the largest and most successful commercial winery prior to Prohibition. It produced thirty thousand to thirty-five thousand gallons of wine in 1900. Twenty thousand of the forty thousand gallons of wine produced by 1902 were shipped to the East Coast. The remainder was sold locally or shipped in horse-drawn wagons to the San Joaquin Valley. The Yorks' story continues in chapter 6.

William (1849–1911) and John (Johan) Ernst (1849–1944): Patriarchs of Seven Generations of Winemakers

Seven generations of the Ernsts and Steinbecks have farmed grapes in Paso Robles. This is the oldest farming dynasty in San Luis Obispo County.

William and Barbara Amelia Ernst were the first to grow grapes and make wine. Their son Frank Ernst farmed with them. William's sons were the first to introduce mechanized farming to the area in the first half of the twentieth century. Their granddaughter Hazel and her husband, George Steinbeck, helped their son Howie and his wife, Bev Steinbeck, plant their

Bev and Howie Steinbeck are the fourth generation of seven generations to farm grapes in Paso Robles. Their daughter Cindy founded Steinbeck Winery on the old Ernst Ranch, where four generations have lived and farmed together since 2009. *Courtesy of Cindy Steinbeck.*

first vineyard in 1982. Howie and Bev's daughter Cindy moved back to the ranch with her children, Ryan and Stacy, in 1997. Cindy's children are both now married and work in the family business. The seventh generation is learning about grapes as they play in the vineyards. The family's love of the land and commitment to farm sustainably has been passed down to each generation. Four generations currently live together on the old Ernst-Steinbeck Ranch on Union Road, which was purchased in 1921 by Frank Ernst to farm the Steinbeck Vineyards. The winery's tasting room features Steinbeck wines and is a family museum of photographs and tools dating back to the life and times of William Ernst.

Twin Legends

William and John Ernst, twins of German ancestry, were born in Baden, Germany, in 1849. Their parents immigrated to the United States prior to 1875, following friends and family members to the German Lutheran community in Geneseo, Illinois, where they farmed and became prosperous. In 1875, Barbara Amelia Mathis married William Ernst. They lived in Geneseo, Illinois, for ten years before deciding to "go west."

William Ernst visited California in the fall of 1884. An announcement in his local Illinois newspaper stated that three thousand acres next to an old Mexican land grant were to be sold in San Luis Obispo County; this land was advertised as good farmland. The announcement asked German Lutherans to join their brethren to set up a community and place of worship. The announcement was signed by Charles Pepmiller. William traveled to explore the area. The local land developers promoted the area in publications such as the *Pacific Rural Press*, stating, "Since the subdivision of the large land grants in this vicinity a new era had dawned, and the region is destined to spring into prominence as a fruit-growing district within a few years." William, once he determined that the opportunity was worth the investment, met with Charles Pepmiller and purchased eighty acres for his new home on a hill north of what is now Creston Road.

William returned to Illinois and shared the news of his investment with his brothers. Both John and Martin decided to move their families to California. William, Barbara and their children left almost immediately and arrived at the Pepmillers' home on Christmas Eve in 1884. John and Martin stayed in Illinois to conduct the property sales for all three families

William and Barbara Amelia Ernst, the first winemakers to grow and produce Champagne in the Paso Robles area, circa 1900. *Courtesy of Cindy Steinbeck.*

and complete the family's business transactions. They sold everything for cash but offered terms for items above ten dollars, according to the signs posted at the auction.

In late 1885, John and Martin Ernst, their families and their in-laws joined William. John purchased the Pepmiller Ranch, and the Pepmiller family moved to a new location nearby. Other families of German Lutheran faith had arrived earlier in the year. They each purchased their own land to farm and built homes in the Creston area. There was a lot of social and religious unity shared in the community. William and John

A sign advertising the sale of the Ernst family property and livestock at a public sale on Tuesday, November 17, 1885, in Geneseo, Illinois. The sale was conducted by John and Martin Ernst. *Courtesy of Cindy Steinbeck.*

helped build the Lutheran church and community center in the Geneseo School District, which was named by John.

Grain and cattle had been the community's main source of income since 1800. The new population planted diverse crops, such as recommended fruit orchards and vineyards. William farmed barley, wheat and oats with a single plow and a mule. Both William and John planted their vineyards and orchards in the Geneseo District in 1885 and 1886, respectively. The vines were most likely purchased from a nursery located somewhere between Fresno and the village of Arroyo Grande. William and Barbara planted vineyards with over twenty-five varieties of grapes. John's vineyard included varieties of Mataro, Carignane, Burger and Rose of Peru planted on nine acres.

The vineyards were profitable. Additional vineyards were planted by both brothers between 1889 and 1903. The source of the vines was the Agricultural Experiment Station east of Paso Robles, which was established by the University of California. It was a valuable resource for everyone farming in the area. John and William Ernst supported the Agricultural Experiment Station financially and with contributions of physical labor from 1889 to 1902. The brothers participated in the research by providing data on local orchards and vineyards to the research scientists who were monitoring soils and crops. William assisted the researchers in preparing their final report by collecting data on all the orchards and vineyards farmed in the Geneseo area. As a result of this research, John and William Ernst concluded that vineyards could be farmed more successfully than orchards on their respective properties. They each reduced their acreage devoted to orchards and increased their acreage of vineyards.

William and John each grew their own grapes but partnered to produce wine under their label, Ernst Brothers. The 1891 Directory of Grape Growers, Wine Makers and Distillers of California shows the Ernst Brothers of Creston as winemakers with eight acres of wine grapes producing Zinfandel wine. They made wine that was considered to be high in quality and equal to the best French and German vintages, according to competent winemakers in Europe. In 1899, Ernst Brothers received orders from those countries, and it was the first winery in the area to do so. In addition to selling their wines to the local trade (bars, restaurants and saloons) in Paso Robles, Ernst Brothers shipped their wines to other areas in the state. In 1900, the local press also reported, "William Ernst will make about 1,500 gallons of wine this year. John Ernst will make about the same and Martin Ernst, 700 gallons. They do not make cheap wines." The first sparkling wine was produced in North County. William and Barbara produced the first Champagne varietal in the area. They won first prize for their Champagne at the California State Fair.

The Ernst Brothers set up a display in 1900 at a local fair, according to a newspaper article that is currently on display at the Paso Robles Pioneer Museum. The article states, "No finer display in the fair can be found than that shown by the Ernst Brothers of Creston." The wine display highlights Claret, White Zinfandel, Champagne, Sparkling Tokay and wine of the Mission grape and dessert wines, including Malaga, Port and Sherry.

The Ernst brothers continued making wines and marketing directly to consumers until William's death in 1910. Multiple generations of Ernsts and Steinbecks have farmed the land, grown grapes and made award-winning wines since 1885.

Gerd Klintworth: First Licensed Winemaker of Jug Wine

Gerd Klintworth had a strong educational and practical farming background in addition to viticulture training. He came to San Luis Obispo County after the vineyards where he worked in Southern California were destroyed by what became known as the Anaheim "blight" or Pierce Disease in the early 1880s. He knew how to select the land with the best soil and climate to farm efficiently. He had experienced a cycle of growth and death in the vineyards of the Anaheim Colony, so he had the experience he needed to successfully farm his new land and vineyards.

His Legend

Gerd Klintworth and his wife, Ilsabe, emigrated from Hanover, Germany, after Gerd's military service was completed (1879–1881). They first settled in Southern California in the renowned Anaheim Colony, which was populated with many people of German heritage. Originally located on the modern site of Disneyland, this colony was "a well-planned, well-executed agricultural experiment devoted to the production of grapes and wine," according to Thomas Pinney in his book *History of Wine in America*.

Gerd moved to California to build a career in the wine industry. He learned about grape production and winemaking in the vineyards of the Anaheim

Left: Labels designed by winemaker Gerd Klintworth, the first licensed winemaker in the Geneseo District, circa 1890. *Courtesy of the Violet Ernst estate.*

Below: The Klintworth family harvesting old vine Zinfandel grapes, circa 1900. *Courtesy of Cindy Steinbeck.*

Colony. The Klintworths moved to San Luis Obispo County after Pierce Disease decimated the vineyards in Southern California. The Klintworths purchased eighty acres next to the Geneseo area, where Gerd built his own two-story home, barn and blacksmith shop and dug a well.

Gerd became well known for the quality of his wine. He planted his vineyards on ten acres. Gerd Klintworth originally planted two acres of orchards and six acres of vines. But, because of the research collected at the UC Agricultural Experiment Station in 1902, he added four acres of vines. Gerd was the first winemaker to be licensed to sell jug wine in North County. The wines that the Wine History Project of San Luis Obispo County can document were made from the grapes grown in Gerd's vineyards: Zinfandel, Claret, Port and Muscatel. The vineyards that Gerd planted with a variety of wine and table grapes still exist on Feenstra Road, just west of Cripple Creek Road.

In addition to his home and blacksmith shop, Gerd raised grain and cattle. He planted orchards with deciduous fruits, including peaches, pears, plums and almonds, which proved to be very successful in 1917. Local agricultural historians credit the Klintworth and Ernst families with establishing grain farming in the Geneseo District. Gerd used a horse to pull his equipment and later described the most significant change in his lifetime: the transition from horse-drawn equipment to large tractors and self-propelled combines.

Gerd had the capital he needed to invest in his wine business. He installed a wine press, which was in use well into the mid-twentieth century. He used a variety of tools, including the wooden fork and the wine press, to crush the grapes; the wine was fermented and stored in wooden barrels with openings plugged with round wooden bungs. Gerd's grandson Robert told the Wine History Project of San Luis Obispo County that the wine that was stored in the barrels in the cellar under the family's home until recently was removed by the new owners in 2020.

Gerd grew both table and wine grapes. He made wine for commercial sale until Prohibition. When Prohibition began, Gerd sold his grapes to local wineries, including the Pesenti and Rotta Wineries in Templeton, and made wine for home consumption. He never reopened his commercial winery. The vineyards were tended until the family home and farm were sold in 1997. The old Klintworth vineyards contain old vines that are still producing quality grapes. One can recognize these sturdy vines by the absence of stakes or trellis wires supporting them.

Gerd and Ilsabe's family included seven children: Henry, Emma, Fred, Chris, Mary, Minna and William. All the children worked on the farm and went to school in the Geneseo Schoolhouse. Their four sons remained in the county and worked in farming until their deaths. Gerd was an active member in the community, joining politics as a Republican, and he served as an elder at the German Lutheran church in Geneseo. The entire family has participated in actively supporting Paso Robles, the El Paso de Robles Pioneer Museum and local friends and family. Gerd retired with Ilsabe to Orange, California, and passed away in 1941 survived by Ilsabe.

Historic Sites to Visit

Epoch Estate Wines (formerly York Brothers Winery)
San Luis Obispo Railroad Museum
Rural communities of Creston and Shandon
Paso Robles Historical Society
Paso Robles Pioneer Museum
Steinbeck Vineyards & Winery

4

The Italians and Old World–Style Farming in the Templeton Gap

GRAPE VARIETIES: FIELD BLEND OF ZINFANDEL, ALICANTE BOUSCHET AND CARIGNANE

Introduction

Italians from Northern Italy and Switzerland had major roles in shaping the wine history of San Luis Obispo County. The wine was a defining beverage in Italian culture and in the sacraments of Catholicism. The bleak economy throughout Italy led to an exodus of over 14 million Italians leaving the country between 1876 and 1914. The majority of these expatriates were young men with no future in their homeland; many came to America to make new lives. There was a higher risk of them dying by staying in Europe than by immigrating to America. Once established, these men sent for their sweethearts and wives to join them.

Northern California was the destination for many of these Italian immigrants in the 1880s and 1890s. Many men in the first wave of immigration in the 1880s and 1890s came from central and southern Italy, including Sicily, where there was an economic and farming crisis. Many southern Italians found work at the Italian Swiss colony in Asti. The Asti Colony and its Italian workers became a national brand that grew Italian varieties and shipped bulk quantities of table wine to every major city with an Italian population. Tipo Chianti wine (*tipo* meaning "type" in Italian) was the most popular. The Italian Swiss colony's brand became the largest table wine producer by 1937. By the 1950s, it was the second-most-visited tourist attraction in California after Disneyland.

Those who migrated to Templeton in the early 1900s had a different heritage. They were typically from the educated middle class from villages in Northern Italy and Switzerland. Many grew up on dairy farms and worked as woodcutters. As they arrived by ship on the East Coast, most of the men became laborers and lived in Italian tenement houses. They worked on the railroads in New York and in the coal mines of Kentucky, Ohio and Pennsylvania. They were paid a dollar a day with gold coins. Eventually, many of these Italian immigrants traveled with friends to California by train, seeking jobs in the dairy, agriculture, lumber and charcoal industries.

The Templeton Gap stretches from York Mountain to Paso Robles in the east. The names of early settlers—Busi, Dellaganna, Dusi, Martinelli, Nerelli, Pesenti and Rotta—are still well known there, more than a century later. Most of these Italian immigrants started as woodcutters and charcoal makers in the early 1900s, clearing the forests to plant vineyards on York Mountain. Charcoal was a valuable commodity; it was used for heating and for the manufacturing of gunpowder.

The Italians moved on to work in the newly planted vineyards. They were trained to dry-farm, prune and harvest grapes. Within a few years, each family had purchased farmland with their savings. The Paso Robles and Templeton areas are similar in geography and climate to Northern Italy; crops and vineyards in both places are planted on rolling hills and steep mountains. Wine grower Frank Nerelli describes the two areas as "so similar that you forget which country you are working in."

These Italian families lived, worked, ate and prayed together, forming strong social bonds. Many had come from the same villages in Italy. Multiple generations often lived and farmed together. They socialized on the weekends at barbeques and card games.

During the 1920s, it was the Italians who transformed the Templeton landscape by removing forests to farm grains and plant Zinfandel in their own small vineyards. They primarily focused on growing grapes; only five Italian growers founded wineries.

In 1920, Prohibition took effect in America, which banned the sale of liquor. A majority of Italians planted their vineyards after Prohibition began in 1920. They produced wine at home and in their wineries for religious and medicinal purposes; a few sold wine to friends and other consumers on the black market. Some were arrested and fined by local law enforcement. Their wine, which was sold by the gallon in jugs with screw tops, was known as "jug" wine until the 1970s, when a new breed of winemakers transitioned Zinfandel from the jug to an elegant bottle.

Multiple Generations of Italian Families Farm in Templeton

The First Generation

Adolphe Siot, the First Viticulturist in Templeton (1858–1925)

The town of Templeton, located within the former Rancho Paso de Robles Mexican Land Grant, was developed by the West Coast Land Company in 1886. For three years, Templeton was the terminus of the Southern Pacific Railroad, bringing immigrants of French, Swedish and Italian origins to settle in the area.

The earliest viticulturist in the area was Adolphe Siot, a Frenchman who bought land from the West Coast Land Company in 1891. His land was located to the west of the town in what is described today as the Templeton Gap, nestled in the Santa Lucia Mountain Range. Siot was the first to plant Zinfandel vines and build his own winery. He was the only commercial winemaker until 1917.

Adolphe Siot Meets Gerome Rotta (1885–1958)

Gerome (Joe) Rotta immigrated with his brother, Clement, to Northern California in 1905; there, he found work in the dairy industry. Joe traveled to San Luis Obispo County to buy land with his savings. He purchased 120 acres of farmland from vineyardist Adolphe Siot in 1908. Joe planted grains and vegetables and raised chickens, livestock and dairy animals. Joe grew up on a dairy farm in the canton of Ticino in Switzerland, but he didn't know anything about viticulture. Adolphe befriended Italian and Swiss immigrants whom he mentored in wine growing and winemaking.

Joe and his wife, Anitta, planted their vineyards on the steep rolling hills in 1919. They were the first Swiss Italians to plant a field blend dominated by Zinfandel in Templeton. The cuttings probably came from Adolphe's vineyard, and the labor was shared between Joe and his mule. Joe soon invited his brother, Clement, along with his wife, Romilda, to join them to work the farm.

Adolphe Siot taught both Joe and Clement his methods for dry-farming; the vines were head-pruned in the Old-World style. Adolphe also taught Clement the art of winemaking. Clement and Romilda were enthusiastic about the vineyards. The vineyard's first vintage was likely produced and consumed by family. The year Adolphe Siot died, 1925, Joe sold his vineyards and farm to his brother, Clement Rotta, and moved on to new ventures.

Lorenzo Nerelli (1883–1968)

Lorenzo Nerelli was born in the province of Foggia, in the region of Puglia, Italy, on January 23, 1883. He was raised on a farm and educated in the Italian public schools. He arrived in New York in 1906 and migrated to San Luis Obispo County in 1915, after working for eight years on the railroads in New York. Lorenzo found employment as a charcoal worker but soon started his own business, hiring his friends as employees. Lorenzo and his crew of workers cleared most of the land around Templeton and Paso Robles. Lorenzo also worked for vineyardists Andrew York and James Robert Anderson on York Mountain. Lorenzo learned to plant, prune and harvest grapevines in their vineyards.

Lorenzo and his new bride, Cesarina Nonini, soon purchased their own 104-acre ranch at the foot of York Mountain. They planted a vineyard and built their own winery in 1917, naming it Templeton Winery. The jug wine they produced was Zinfandel. During the Prohibition years, they continued growing grapes and producing limited quantities of wine. Templeton Winery was the second winery in San Luis Obispo County bonded after Prohibition was repealed in 1934. The *San Luis Obispo Telegram* reported on page 1 on April 24, 1934, "Winery Going Up Near Templeton." Lorenzo hired Joe L. Forti of Fresno to build a large winery on his ranch with cement brick brought from the valley. Lorenzo made high-quality wines, which he sold in bulk to wineries in Northern California that were shipping to eastern markets.

President Franklin Delano Roosevelt declared war on Japan after the bombing of Pearl Harbor in 1941. The sons of the Italian families were eager to enlist. Lorenzo Nerelli was forced to close his winery when all three of his sons left to fight in World War II. Unable to sell the property, Lorenzo and Cesarina abandoned their York Mountain ranch, including the vineyards and the winery. They moved to Paso Robles to purchase sixteen acres of land on Pacific Street, where they planted a new vineyard. Lorenzo continued to grow grapes, selling his annual harvest to Pesenti Winery. He lived in Paso Robles until his death in 1968.

Giovan B. Busi: Charcoal and Grapes on York Mountain (1886–1939)

Giovan "Bob" Busi settled in San Luis Obispo County in 1914, after working the gold mines in Jackson, California. Bob originally emigrated from a village in Northern Italy; afterward, he found work clearing land and making charcoal in the Santa Lucia Mountain Range. Bob landed a contract with John Pesenti, the first of the Pesenti family to settle in Paso Robles in 1907. Bob hired a crew but needed additional workers. He invited his brothers to join him.

A photograph of Sylvester Dusi and Giuseppe Busi clearing the land and making charcoal in Templeton, circa 1921. *Courtesy of Mike and Joni Dusi.*

The three Busi brothers lived in the mountains in a "wood cutters camp"; food and supplies were delivered by John Pesenti. They also worked with Lorenzo Nerelli on his ranch on York Mountain. In 1920, the Busi brothers purchased sixty-three acres of land to farm, and they planted grapes next to the Nerelli ranch. The Busi family became known as experts on charcoal production.

Frank Pesenti (1896–1983)

Frank Pesenti and his sister traveled to San Luis Obispo County from their small village of Brembilla in Northern Italy with Bob Busi's brothers in 1914. At the age of eleven, Frank started training as a mason's apprentice in France. He worked there for seven months each year, returning to Brembilla each winter. At the age of eighteen, Frank decided to follow his uncle to San Luis Obispo County.

Frank worked as a charcoal maker for Giovan Busi, saving his earnings to purchase the Ward Ranch in the Templeton area in 1919. He sent for his wife, Caterina, who arrived from Italy in 1922, and they planted Zinfandel vines on the steep hills the same year.

Frank was making wine within three years, crushing grapes and letting them ferment naturally in redwood barrels. Redwood was the most plentiful pliable wood available for barrel making in California. Traditional Italian winemaking did not include aging in oak barrels or focusing on quality. In fact, most wines were consumed within a year of production. During the Prohibition years (1920–1933), Frank sold grapes to the Basque population in central California. He also made two hundred gallons of wine per year, the legal limit per household.

Frank used his skills as a mason to manufacture his own bricks to build his winery. His son Victor laid out the molds for production. Pesenti Winery was the first to be bonded in San Luis Obispo County in 1934, following the repeal of Prohibition. It was located on Vineyard Drive in Templeton, next to the fields and vineyards of the Rotta family. Frank was the winery's winemaker from 1934 to 1946.

Frank and Caterina raised five daughters and one son. Like many of his classmates, their son Victor Pesenti enlisted in the army in 1943, and he was assigned to the Mountain Infantry Division. Both Victor and his sister Silvia married their spouses in 1946. Victor joined his father in the wine business immediately afterward; Silvia and her husband, Aldo Nerelli, joined the

Left: Frank Pesenti, the founder of the first winery that was bonded after Prohibition in 1934, Pesenti Winery. *Courtesy of Frank Nerelli.*

Right: Victor Pesenti, the son of Frank Pesenti, standing near Zinfandel vines in the Pesenti Vineyard, which was planted in 1922. *Courtesy of the Paso Robles Historical Society.*

winery staff in 1948. The second generation soon moved into winemaking and the management of Pesenti Winery. Aldo and Victor worked together as the winemakers from 1946 to 1969.

Pesenti Winery was famous for Zinfandel and generic red table wines sold in gallon jugs. The winery's grapes were sourced from their estate vineyards. As the business expanded, additional Zinfandel grapes were purchased from Amedeo Martinelli and other local Italian growers in Templeton. Pinot Noir and Cabernet Sauvignon were added in the 1960s, and they were also available in jugs and bottles.

Sylvester (1883–1964) and Caterina Dusi (1894–1985)

The four Dusi brothers from the village of Ono Degno, in the province of Brescia, Italy, immigrated to the United States at the turn of the twentieth century. They grew up in Northern Italy on their family's dairy farm. Lorenzo was the first to arrive in 1903, joining his uncle in Oregon. By 1904, Lorenzo had moved to San Luis Obispo County and was working at Mainini Dairy. Lorenzo had come to America to escape the dairy business, so he chose to work as a woodcutter, clearing trees for Lorenzo Nerelli on York Mountain. In 1905, Joe Dusi joined Lorenzo after working in the Pittsburgh mines. Sylvester, the eldest, arrived in California in 1907. Sylvester sent for the youngest, Dan, after visiting Lorenzo in 1920. By 1921, all four brothers were working for Lorenzo Nerelli, clearing trees. The three younger Dusi brothers purchased a 227-acre ranch at the top of York Mountain in the early 1920s. There, they planted grapes and sold them to York Brothers Winery.

Sylvester started looking for new opportunities in Paso Robles. He was thirty-eight years old with an eye on the future. Anticipating the future growth in San Luis Obispo and Paso Robles, he purchased the Brendan House at 1122 Pine Street in Paso Robles with a partner in 1921. Sylvester converted the second floor of the home into a boardinghouse, renaming it the Hotel d' Italia. On the street level, he opened an Italian restaurant and grocery store. His business partner was replaced by Caterina Gazzaroli, who traveled from Italy to work at the hotel. They were married in San Francisco in 1922, forming a legendary business partnership. Over the next seventy years, they owned three hotels, three restaurants, three legendary vineyards, a gas station, a liquor store, a billiard parlor, two wineries and a sausage company. Sylvester identified these business opportunities, and Caterina's management made them profitable.

The first of the couple's three sons, Guido, was born on the second floor of the hotel, where the family lived. In 1924, during Prohibition, Sylvester saw the potential in growing grapes for home winemaking customers. Sylvester and Caterina purchased their first ranch of 87.9 acres in 1925, shortly after their son Dante was born. In 1926, Sylvester hired Steve Medrano to help clear the land and plant their vineyard with a traditional Italian field blend; the planting of the vineyard was completed in 1927. The majority of vines were Zinfandel, but there were also small amounts of Carignane, Mission, Alicante Bouschet, Petite Sirah and Grenache grapevines. The vineyard is now known as the Benito Dusi Vineyard, named for the couple's youngest son, who was born in 1933.

Today, the Dusi family is famous for growing Zinfandel grapes that are among the best in California. After Prohibition ended, Dusi grapes were sold directly to winemakers; the Rotta Winery became a major customer between 1938 and 1945.

When the Dusis purchased the property, which came to be known as the Dusi Ranch, Sylvester and Caterina built their home there. It had an

Top: Sylvester Dusi holding his newborn son, Guido, in front of the Hotel d' Italia on Pine Street in Paso Robles in 1923. *Courtesy of Mike and Joni Dusi.*

Bottom: Benito Dusi on the "Cat 10," tractor farming the Zinfandel vineyard on the home ranch now known as the Benito Dusi Vineyard, circa 1944. *Courtesy of Mike and Joni Dusi.*

Opposite: Clearing the land and making charcoal in 1946 at the second vineyard, now known as the Dante Dusi Vineyard. Sylvester Dusi can be seen seated on the truck, Dante is holding the handle and the child on the crawler is Benito. *Courtesy of Mike and Joni Dusi.*

extraordinary kitchen in the basement, where Caterina cooked polenta and stew every Sunday for multiple generations. The couple raised their three sons, Guido, Dante and Benito, in the Dusi vineyards and in San Luis Obispo. The family moved to San Luis Obispo in the early 1930s so that Guido and Dante could attend the Mission Catholic School. Sylvester bartered wine to pay their tuition, and the church insisted on receiving the Dusi Port. In 1931, the Dusis purchased a gas station, a brick building and a liquor store in the city of San Luis Obispo. Caterina opened the La Pergola Liquor Store on January 11, 1936, selling local wines from Pesenti and York Brothers Wineries with a five-gallon limit. Dante made the deliveries to the madams in the red-light district, which he truly enjoyed.

By 1938, Caterina was cooking the meals at the Italian restaurant and managing Stag's Billiard Parlour. Sylvester, who was affectionately known as the "Godfather," was generous and supportive of those who were less fortunate. He gave away his sausages to those who were unable to pay. He bought cars and farm equipment from those dealers on the verge of financial ruin to keep them in business.

During the Great Depression in the 1930s, the demand for grapes tapered off. Sylvester and Joe Dusi teamed up with twelve other Italian Zinfandel growers, including the famous pianist Ignace Paderewski, to take out advertisements in the local newspaper. Their grapes were sold at twenty-eight dollars per ton in 1936. However, grape prices continued to decline.

Sylvester decided to finance a winery in 1938 so that growers could make bulk wine with the surplus of grapes. The San Luis Winery was located in downtown San Luis Obispo. Sylvester bought new winemaking equipment and redwood barrels. Unfortunately, those barrels were not treated, and the wine became contaminated. The ever-resourceful Sylvester was able to sell the wine to a distiller to make brandy. Only one vintage was crushed at the San Luis Winery.

In 1939, the family moved back to the Dusi Ranch and focused on the grape growing business. Benito learned to farm, driving the tractors and tending the grapes with his brothers. By the age of six, he was working on the land that became his home for the next eighty years. Dante and Benito were the second generation of grape growers in the Dusi family.

Amedeo Martinelli (1881–1961)

Amedeo Martinelli was born in the town of Giulianova, in the province of Teramo, Italy, in 1881, and he immigrated to the Templeton area around 1920. He joined fellow Italians working in the charcoal industry after World War I. He bought his own property, twenty-five acres on Ridge Road in Templeton, in 1926.

Following Italian tradition, Amedeo planted cherry trees among the Zinfandel vines in his vineyard. He built his own winery and was well known for producing fine wine, which he sold locally. His cherries were also highly prized. Amedeo was a popular man and a talented winemaker. He was self-assured and full of laughter. He enjoyed playing cards and gambling with his dear friends Joe Dusi and Roy Giandinni.

Many years later, while on a trip to Posina, Italy, in 1958, Amedeo met and fell in love with the widowed Rina Fossati, a mother of five children. Amedeo had no children of his own and wanted to pass his vineyard to a second generation. Amedeo had recently won a restaurant in a game of cards, so he decided to propose to Rina and invite her and her children to join him in Templeton. When she accepted, he returned home and sold

Young Bruno Martinelli (*center*) surrounded by his family in Posina, Italy, circa 1959. The family reunion occurred just before his mother, Rina, married Amedeo Martinelli, who founded his winery and historic vineyard in Templeton in 1926. Bruno was adopted by Amedeo. *Courtesy of Bruno Martinelli.*

the restaurant to gain some cash. He quickly returned to Posina, where Amedeo and Rina were married by a judge. He adopted two of Rina's daughters, Marie and Silvana, and one of her sons, Bruno. Amedeo, Silvana and Rina flew to California in 1959; Marie and Bruno arrived by plane on Christmas Eve in 1960.

Amedeo died from cancer a few weeks later, on January 16, 1961. It was a terrible shock. Rina and the children spoke little English and knew nothing about growing grapes or making wine. Amedeo's dear friends Joe Dusi and Roy Giandinni helped Rina and Bruno learn the vineyard business. They eventually closed the commercial winery. The vineyard was well-established and produced premium Zinfandel grapes. Pesenti Winery had a long history of purchasing the vineyard's harvest. Joe and Roy trained Rina and Bruno to prune and harvest the vineyards. All three children worked alongside their mother in the vineyard before and after school and all throughout the weekends. It was difficult work for twelve-year-old Bruno, but he loved the grapes and the cherries. Amedeo's legacy was preserved by Bruno for four decades.

The Second Generation

Rotta Winery Becomes Legendary for Jug Wine and Three Generations of Winemakers

Clement (1890–1963) and his wife Romilda (1895–1976) expanded the Rotta vineyards and built their own winery after purchasing the farm from his brother, Joe. They were the perfect team for the next fifty years. During the Prohibition years, Clement sold his grapes and juice for home winemaking; he also made wine for the Catholic church, possibly Mission San Miguel. Clement planned to make wine commercially. Rotta Winery was the third in San Luis Obispo County to become bonded after Prohibition ended in 1934.

Clement and Romilda built their own winery in 1937. They bought a ratchet-operated grape press and several eight-thousand-gallon redwood tanks for aging. Concrete tanks were installed for fermentation. A large supply of gallon jugs and screw tops were used for bottling. Their Zinfandel was rustic and very reasonable in price. They always sold directly to the public at all times of the day—there were no official hours.

Rotta Winery, known for its unique Central Coast architectural style, made jug wine, the favorite of California Polytechnic State University students, surfers and members of the hippie generation, circa 1940. *Courtesy of Mike Giubbini.*

Mervin Rotta, a ranch hand, Bob Giubbini, Joe Rotta and his brother and winemaker Clement Rotta sampling the fruits of their labor: Zinfandel jug wine. *Courtesy of Mike Giubbini.*

Rotta Winery developed a loyal following of locals who brought their own unique containers (now highly collectible), which Romilda filled by ladling Zinfandel into them from a redwood cask. If she wasn't available, visitors could fill their own gallon jugs and leave $2.25 in the cigar box on top of the barrel.

Clement made four types of Zinfandel, including a sweet dessert wine; the "old Zinfandel" was aged for seventeen years in the redwood tanks. Eventually, the Rottas added a tasting room to the winery that was made from an old seven-thousand-gallon redwood tank. Clement passed away in 1963. His son Mervyn joined Romilda as the second generation of Rottas to work at Rotta Winery.

Aldo Bruno Nerelli (1917–2007) and Silvia Pesenti (1923–2018)

Aldo was born on York Mountain on April 6, 1917, to Lorenzo and Cesarina Nerelli. He worked with his family in the vineyards starting in early childhood, and he worked at the Templeton Winery until he enlisted as a soldier in World War II.

When Aldo returned in 1945, he found the Templeton Winery and vineyards abandoned. However, Aldo continued the winemaking tradition, the second generation in the Nerelli family to do so. He married Silvia Pesenti, whose father, Frank, founded Pesenti Winery in 1934. Silvia was born to Caterina and Frank Pesenti in 1923. The youngest of six, Victor, was the only son. The Pesenti children worked in the vineyards and the winery with their parents. Silvia and Vic only spoke Italian when they entered the Templeton School.

While Aldo was abroad during World War II, Silvia worked just north of Paso Robles at Camp Roberts, which was built in 1940 as a World War II training center. They were married in 1946 and remained happily together for over sixty years. Aldo's first job was in the liquor distribution business in San Luis Obispo. Two years later, Silvia's father invited the couple to move to the family farm. Aldo joined the staff of Pesenti Winery in 1948; by 1950, they had built their own home at the top of the vineyard, where they raised three children, Frank, Ann and Terisa. Silvia worked in the winery for over fifty years, cooking lunch every day for her family, friends and workers. Her gnocchi and polenta were always in high demand.

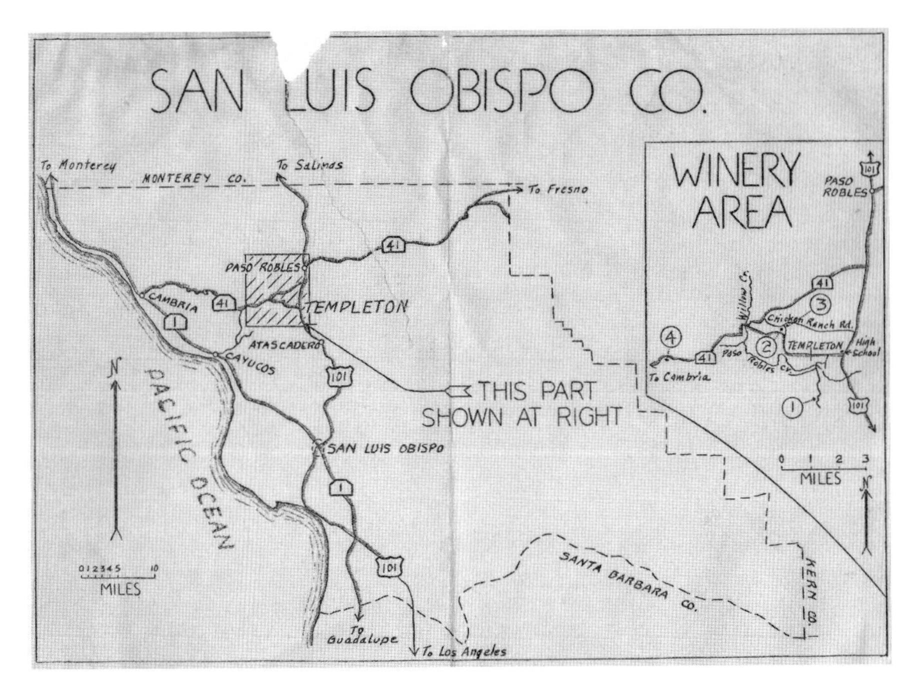

A San Luis Obispo County map, dated 1945, showing the location of the four wineries in Templeton: Martinelli, Pesenti, Rotta and York Wineries. *Courtesy of Suzanne Goldman Redberg.*

Aldo Nerelli (1917–2007) and Victor Pesenti (1924–2000)

By 1970, only three local wineries had been in business for more than thirty years in San Luis Obispo County: York, Rotta and Pesenti. The York and Rotta families grew their own grapes and made Zinfandel wines, focusing on local customers and tourists. The Pesenti Winery grew its own grapes and purchased grapes from many small Italian growers, including Lorenzo Nerelli and Bruno Martinelli. Pesenti Winery sold to customers statewide, including the Basque population in the Central Valley, the Swiss dairy farmers along the Central Coast, sheepherders in North County, restaurants, liquor stores and wine distributors. Victor Pesenti and his wife, Linda, joined the Pesenti family at the winery in 1946, when Victor returned from World War II.

Victor and Aldo worked together until Victor died in 2000. The winery was sold to Larry Turley in 2001. They worked as a team, making Pesenti wines from 1948 to 1969. Frank Nerelli, Frank Pesenti's grandson, was appointed winemaker in 1970. Victor was the sociable, outgoing face of

An aerial view of the Pesenti Winery, located in the rolling hills of Templeton. *Courtesy of Frank Nerelli.*

the winery, marketing Pesenti wines throughout California. Aldo was the businessman who ran the production and distribution of the wines. He hosted tastings and tours at the winery. They even expanded the varietals; Zinfandel Rosé was the favorite, according to Aldo. Frank Nerelli produced a four-star Port, which was highly praised.

Pesenti Vineyards were famous for their old vines and the quality of their grapes. Pesenti jug wines continued to be favorites all throughout California. The California food revolution was focused on crafting premium wines, and many winemakers sought to buy Pesenti grapes.

Guido (1924–1997), Dante (1925–2014) and Benito Dusi (1933–2019)

As soon as they graduated from high school, the Dusi brothers enlisted to fight in World War II. Guido enlisted in the navy in 1942, and Dante joined the U.S. Coast Guard in 1943.

Caterina and Sylvester purchased a second ranch in Templeton in 1945, when the war was coming to a close. They cleared the land and made charcoal with the help of Giuseppe Busi before planting Zinfandel, Alicante Bouschet and Carignane grapes. Benito drove the vineyard's Caterpillar tractor and helped farm the Dusi vineyards. Benito loved grape growing; he farmed the original Dusi Ranch until his death in 2019, although he also worked day jobs for many of those years.

Guido and Dante returned to San Luis Obispo County in 1946 to plant the new vineyard. Guido married Earnestine Dobbs in 1950; he chose to open his own electrical business in Paso Robles. Dante married Dorothy Steppie at the San Antonio Mission in 1949. Dante worked six days a week for Madonna Construction for the next three decades, devoting his weekends and evenings to farming. Dante and Dorothy built their home overlooking the vineyard; their children, the third generation, Rick, Mike and Kathy, worked with the family after school and on the weekends. Dottie, a "city gal" from Chicago, managed the harvest crews at both vineyards while working at the telephone company in the evenings. Dusi grapes were in high demand; they were sold to York Brothers Winery from 1945 to 1960, Bertero from 1960 to 1965 and Paul Masson from 1965 to 1973.

This second Dusi vineyard was known as the Dante Dusi Vineyard. Dante's son Mike enjoyed working in agriculture, farming grain near Bradley for thirty years. Mike and his wife, Joni, built their home and raised their children at the Dante Dusi Vineyard. Since 1972, Joni had farmed the vineyards with Dante while raising the fourth generation of Dusis, Michael, Matt and Janell, just steps away from their grandparents.

Janell spent much of her childhood with her grandfather, working with him in the vineyard. By the age of twelve, she was questioning him about why all the Dusi grapes were sold to famous winemakers instead of being used to make Dusi wine. Dante helped her make her first vintage of Zinfandel in 1992. She won honorable mention at the California Mid-State Fair in 1993. In 2006, she founded her own label, J. Dusi, which was printed in Dante's favorite color, "Dusi blue." She has been making Zinfandel and new varietals at her own winery since 2013, with Dante's blue truck sitting out front to welcome guests. Sylvester and Benito each made wine with their surplus grapes. The wine was sold through the tasting room located in a small building behind their home on the Dusi Ranch from 1954 to 1967. Janell is the first in four generations of the Dusi family to be totally devoted to making wine.

The Dusi family standing in the Dante Dusi Vineyard, just south of Mike Dusi's home. The third and fourth generations of the family in 1993 (*from left to right*): Matt, Janell, Mike, Joni and Michael. *Courtesy of Mike and Joni Dusi.*

Caterina purchased a third ranch in 1945, and it became known as Caterina's Vineyard. Fourth-generation member Matt and his wife, Ali Dusi, purchased the property in 2002. They planted Syrah grapes on eleven acres. Matthew traveled to France to learn farming techniques and trellis styles for this new Rhone grape variety. Dante and Joni Dusi planted the remaining eleven acres with Zinfandel.

Michael Dusi's son, Dante, is a part of the fifth generation of Dusis to farm the iconic vineyards.

Historic Sites to Visit

J. Dusi Winery
Rotta Winery
Turley Wine Cellars
Zin Alley, owned by Frank Nerelli

Scenic Highway 46 West, from Highway 101 to the Pacific Ocean
Camp Roberts and the Camp Roberts Historical Museum, the largest army museum in California
The Museum Annex at Camp Roberts, a collection of historic army vehicles

5
Prohibition

Before and After (1900–1960)

GRAPE CHOICE FOR PROHIBITION: WINE BRICK FOR HOME WINEMAKING

Introduction

Between 1900 and 1960, five major events shaped the wine history of San Luis Obispo County: World War I, the Prohibition era, the emergence of home winemaking, World War II and the continuity of small family-owned vineyards and wineries.

During World War I, local agriculture shifted from growing grapes and grain to growing beans that could be dried and shipped abroad to the starving populations and soldiers in Europe. San Luis Obispo County was listed as one of the six leading counties in California for bean production, according to the "California Statistical Report of 1910." At the same time, the local wine industry, which was then led by the York Brothers Winery, rapidly expanded its production.

The Eighteenth Amendment established the era now known as Prohibition and instituted a nationwide constitutional ban on the production, importation, transportation and sale of alcoholic beverages between 1920 and 1933. It was supported by a patriotic campaign during World War I.

This amendment was passed by the United States Congress and was ratified by forty-six states. The amendment did not make it illegal to drink, however; any wine, beer or spirits in the possession of an American could be enjoyed and consumed at home. It penalized the producer, not the consumer. The Volstead Act, which was written by Congress to define the implementation and enforcement of the Eighteenth Amendment, stipulated that individual states should enforce Prohibition. Therefore, Prohibition enforcement and history varies by state.

Most California winemakers were in denial about the possibility of the passage of Prohibition, even as the Eighteenth Amendment was ratified state by state. An intoxicating liquor was defined as having an alcohol content of 0.5 percent or more. Many had thought whiskey and spirits might be banned but not wine; grapes and wine were viewed as food products.

Pre-Prohibition (1900–1920)

Andrew (1833–1913), Walter (1871–1952) and Silas York (1882–1967)

In 1906, Andrew, Walter and Silas decided to enlarge York & Sons Winery. They purchased additional land and planted new Zinfandel vineyards on eight acres. Small plantings of Grenache, Carignane and Alicante Bouschet were also added. They expanded the winery using bricks due to a shortage of lumber in the county. Andrew hired a local brickmaker to teach the family how to build a kiln and mold the bricks from the clay soil on York Mountain. According to some reports, over one hundred thousand bricks were manufactured on site by family members. Wood for both the beams and the ceiling of the winery, which was salvaged from the old Cayucos pier and a bridge near Jack Creek, was driven to the site by wagon from the Pacific Coastline.

The winery addition was completed in 1907 at a total cost of $3,000, according to Andrew's grandson Sid York. Sid described the new equipment purchased as "7,000-gallon tanks with a cost between $120 and $175 each, 5,000-gallon tanks purchased for around $75 each." Andrew also bought a gasoline engine and a new crusher, which was used at the winery until 1970. Andrew York called his grape harvest of 1907 "the best season in twenty years."

Around 1910, York & Sons Winery was awarded the largest contract ever negotiated in San Luis Obispo County. A San Francisco firm purchased large quantities of Zinfandel wine in bulk, dictating specific directions on how the wine was to be stored, aged and transferred from the winery to the trains at the Templeton Depot to maintain the quality and stability of the wine. In addition to this contract, approximately 50 percent of York & Sons' wine production was shipped to the East Coast.

Andrew York died on December 1, 1913. As their father's health had declined in 1911, Walter and Silas took control of the winery and grew its production of wine to over one hundred thousand gallons annually. Walter York and Silas York purchased their interest from their father's estate and changed the winery's name to York Brothers. The winery had been passed on to the second generation, and York Brothers was then the largest winery in the county—its future looked bright.

Throughout Prohibition, the York Brothers Winery found a market for its Zinfandel grapes. The demand was fueled by two factors: the growing illegal market for wine, which was controlled to a great extent by organized crime, and the home winemaking industry. York Brothers focused on the local markets and continued to sell and press grapes on site, both for Swiss-Italian dairy families and the Basque ranching populations throughout coastal and central California. They also pressed grapes for local winemakers, including the famous Polish pianist and statesman Ignacy Jan Paderewski.

Because of the ambiguity of the Eighteenth Amendment, York Brothers Winery was allowed to sell wine for both sacramental use in religious services and medicinal use in prescriptions. In fact, physicians were allowed to write up to fifty prescriptions for alcohol per month for acute and chronic illnesses.

The Casteel Family: Clarence (1879–1953) and Melvin (1914–1980)

One of the earliest families to settle in the area west of Paso Robles was the Casteel family, who moved to the area in 1887. Jacob Israel Casteel was a Mormon who originally settled on the San Bernardino Ranch in Southern California to establish a city patterned after Salt Lake City, Utah. When the project failed, Jacob's son John Wesley Casteel moved his family to Arroyo Grande in San Luis Obispo County to farm beans. In 1887, they moved to Dover Canyon, west of Paso Robles. Four generations of the family worked in cattle ranching, dairy farming, grain production, tanning buckskin, trapping and selling small animal pelts. But most importantly, the family was growing Zinfandel grapes. Jacob's grandchildren were Clarence and Mattie Casteel. Clarence married Ethel Blanch Heaton on October 2, 1908. They raised four children: Harlan, twins Elvin and Melvyn and Louesa Emily.

Clarence Casteel bought his 160-acre ranch on Jensen Road, northeast of Vineyard Drive, in 1912. There was a vineyard with old Mission grapes planted on the property. Clarence planted 30 acres with two new varieties, which included Muscat grapes and Zinfandel grapes, now known as the heritage grape of San Luis Obispo County. It is thought to be one of the oldest Zinfandel clones in California. During Prohibition, in 1927, Clarence planted additional grapes on 7 acres, and his vineyard acreage continued to increase each year. His grapes attracted a famous customer, media mogul William Randolph Hearst. Hearst wanted the very best table grapes grown in San Luis Obispo County to serve to his guests at his ranch, which was then known as La Cuesta Encantada, located high on the hill overlooking the Pacific Ocean and San Simeon. Clarence's children described the silver coins that he would bring home in buckskin bags from the now-famous Hearst Castle. They "delighted in watching their father pour the silver into a dishpan for counting."

In 1933, Mel and Elvin Casteel, Clarence's twin sons, planted a small vineyard of Zinfandel grapes on seven acres on their father's ranch as their project for the Future Farmers of America. This project inspired Mel to pursue his career as a vineyardist. After Clarence's death in 1953, the ranch was divided, and some of the acreage was sold. Harlan and Elvin had their own farms in Paso Robles, so Mel was able to purchase their shares. Mel moved his family to Clarence's vineyards in 1956. Although Mel pursued a career in the meat business in 1937, working with Bryan Meat Company for over thirty years, he became well known as a Zinfandel grower; his grapes were famous for their rich flavor. He expanded his vineyards and even built a wine cave, where vineyard workers and their families could enjoy relaxing after a long day in the vineyard.

As Mel expanded his vineyards in the 1950s, the huge rocks that were removed from his property became legendary. The boulders were hauled away by Phil Madonna to build the Madonna Inn, an iconic hotel in San Luis Obispo. His twin brother, Elvin Casteel, was the stonemason who designed the rockwork at the Madonna Inn, the San Luis Obispo Mission Plaza and the Paso Robles Fairgrounds.

When home winemaking became the trend in the late 1960s, Mel invited amateur home winemakers from Southern California to harvest their own

Mel Casteel and his neighbor George Dellaganna recording the last harvest of Zinfandel at the Casteel Ranch in 1973. The land, which had a small vineyard of Mission grapes, was purchased in 1912 by Mel's father, Clarence Casteel. *Courtesy of Doug Casteel.*

grapes in his vineyards and enjoy a wild weekend of "Zin and Bar-B-Que." These harvest weekends encouraged many amateurs to become professionals, including Bill Greenough, who later purchased the vineyards in 1974 that had originally been planted by Henry Ditmas in the Upper Arroyo Grande Valley and founded Saucelito Canyon Winery. Mel also sold his grapes to famous winemakers in Northern California, including David Bennion of Ridge Vineyards. Mel mentored many new wine growers planting their first vineyards in Paso Robles during the 1960s. When Mel sold his vineyards in 1972, he continued consulting with local growers on planting layouts and the rooting of vineyards.

Ignacy Paderewski: Composer, Pianist and Viticulturist (1860–1941)

A celebrity arrived in Paso Robles in January 1914, a propitious event for the York Brothers Winery. He was a world-famous pianist, composer and Polish patriot living in Switzerland who suffered pain and stiffness in his hands during a California concert tour. Ignacy Paderewski was advised to postpone his tour and "take the healing waters in Paso Robles" as a cure. He resided at the famous El Paso de Robles Hotel (now the Paso Robles Inn) and sought treatment at the mineral hot springs.

Paderewski's doctor, who also sold real estate, pressured Paderewski to purchase a 2,500-acre ranch in the Adelaida District, west of town, which he did by March 1914. He hired a Polish patriot to manage the ranch, planting orchards of fruit and nut trees at his San Ignacio Ranch. Paderewski resumed his concert tour and continually lobbied for the Polish people wherever he traveled. He worked with Allied leaders to ensure Poland's independence in 1919 and served as the first prime minister of the newly independent Poland.

Paderewski returned to the El Paso de Robles Hotel in 1922 during the Prohibition era. Paderewski had been famously known for the luscious large table grapes he had grown in greenhouses on his estate in Switzerland, Riond-Bosson. Paderewski knew that both the demand and price of grapes were rising. He decided to plant Zinfandel after consulting with two professors from UC Davis, F.T. Bioletti and Horatio Stoll. In 1923, his ranch manager, J. Gnierciah, cleared two hundred acres and planted Zinfandel vines that had been purchased from a nursery in Riverside, along with small amounts of Béclan and Petite Sirah grapes; they were all

Ignacy Jan Paderewski, a composer and classical pianist, planted Zinfandel vines in Paso Robles during Prohibition. The grapes were pressed and made into wine at the York Brothers Winery, winning a gold medal in 1934 at the California State Fair. *Courtesy of the Wine History Project of San Luis Obispo County.*

planted on their own roots. The first harvest was in 1926, and the grapes were sold to locals and the Swiss-Italian dairymen in the Salinas Valley throughout the Prohibition era.

Paderewski hired York Brothers Winery to press his grapes and make his wine at the end of Prohibition. Paderewski entered his wines in the 1934 competitions at the California State Fair, winning gold medals and bringing fame to York Brothers Winery and Paso Robles.

Paderewski died in 1941. His estate was litigated for a decade and then liquidated to pay his taxes and his debts. Some of the remaining land where Paderewski's vineyards were first established is now owned by Epoch Estate Wines.

The Real Story Behind Prohibition in California

The contrast between California's reaction to the evils of alcohol and laws designed to restrict the sale and manufacture of liquor and the rest of the nation's reaction is remarkable. The Eighteenth Amendment was the culmination of almost a century of growth in the temperance movement and religious revivalism that swept through the United States starting in 1830. At this time, California was experiencing the desecularization of the Spanish missions under Mexican rule (1821–1848), followed by the transfer of the territory to the United States.

Kansas was the first state to become "dry," establishing a statewide ban on the manufacture and sale of alcohol in 1880. In the same year, the California State Board of Viticultural Commissioners was formed to educate and support the wine industry. The board published papers on wine grape varieties to plant and quarantine rules to limit the spread of phylloxera. Conventions were held to bring growers and winemakers together to discuss issues ranging from labor to selecting wine barrels.

As the move toward prohibition grew, California voters rejected the Harris State Enforcement Act in four state elections between 1914 and 1920, as they enjoyed jugs of California wine. In 1919, the California Legislature did finally ratify both the National Prohibition Enforcement Act and the Harris Act. However, there were four important loopholes: California allowed the sale of grapes; the state did not prohibit the process of converting juice into wine; it was not illegal to drink intoxicating liquors (beer and wine); and home winemakers were allowed to produce two hundred gallons per household for their own consumption.

Many Californians chose to break the law and purchase alcoholic beverages illegally through bootleggers, local wineries, the black market and the many speakeasies in San Luis Obispo County. Wine production in San Luis Obispo County increased during Prohibition.

Prohibition was repealed by the Twenty-First Amendment in 1933, with the acknowledgment that it had led to an increase in local and federal corruption and the rise of organized crime.

One Hundred Years of Home Winemaking Shapes California and Local Wine History

Home winemaking has thrived in California since the Spanish claimed Alta California. Over the last one hundred years, settlers from around the world have brought their own winemaking skills to California. According to wine historian Charles L. Sullivan, families of French, German, Italian and central European origin used as much as 5 percent of the state's wine grape crops to make wine at home prior to Prohibition. As immigrants settled in cities across the nation in the early twentieth century, wine grapes were shipped to cities for sale for home winemaking. In 1913, hundreds of carloads of wine grapes were shipped from California; by 1917, the number of carloads had risen to four thousand. Sullivan also reported that the number of carloads had escalated to fifty-five thousand by 1923.

It seemed that Prohibition had given rise to a new hobby: home winemaking. The demand for grapes increased due to a loophole in the legislation that allowed each household to make no more than two hundred gallons of wine per year for their own consumption. Home winemaking kits were manufactured and promoted throughout the country during Prohibition. Books, wine bricks, grape concentrate and raisin cakes were sold in every city. The San Francisco Bay Area alone used at least two thousand carloads of grapes for home winemaking each year during Prohibition. And as the demand for grapes increased, so did the vineyard acreage in San Luis Obispo County.

Corkscrews and other bar tools were designed throughout the Prohibition era. Many appeared to resemble Senator Volstead, who is sometimes called the "Father of Prohibition." *Courtesy of the Wine History Project of San Luis Obispo County.*

The demand for red grapes, specifically Alicante Bouschet, Zinfandel, Petite Sirah, Carignane and Mourvedre, grew because of home winemaking. These hardy grapes were preferred for shipping by truck and by rail. The consequences of this trend would be felt in the California wine industry for decades; many fine grape varieties, both red and white, disappeared from production during the 1920s as growers removed the vines and replaced them with grape varieties in demand.

Home winemaking became popular again in the late 1960s. The Home Winemaking Shop was opened by winemaker John Daume in Woodland Hills around 1970. Daume offered classes, seminars with new California winemakers and field trips to UC Davis for amateur winemakers. He also organized the Cellarmasters of Los Angeles Home Winemaking Club in 1974, and it still meets monthly for wine tastings and competitions. Over fifty members of the Cellar Masters have established their own wineries in California.

In San Luis Obispo County, these home winemakers who established their own wineries included: Ron Bergstrom and John Scott of Ranchita Oaks Winery, Doug and Nancy Beckett of Peachy Canyon, Dave Caparone, Hank Donatoni, Robert and Joanne Dunning, Pat Mastantuono, Pat Wheeler of Tobias Vineyard, Andrew and Cathy MacGregor, John Pierre Wolff of Wolff Vineyards and Romeo and Margaret Zuech of Piedra Creek Winery, just to name a few. Two of San Luis Obispo's legendary commercial winemakers also started as home winemakers: Dr. Stanley Hoffman of HMR and Tom Myers of Estrella River Winery and Castoro Cellars.

All growers sold their grapes to home winemakers. Benito Dusi and Mel Casteel continued to do so with the home winemaking generation of the 1960s and 1970s. Benito Dusi had an exclusive contract with Ridge Winery for his Zinfandel grapes, but he always saved around 5 percent of his crop for the John Daume Home Winemaking Shop and locals in San Luis Obispo County.

San Luis Obispo County: Bootlegging and Outfoxing Local Law Enforcement

In San Luis Obispo County, York Brothers Winery sold grapes to local Swiss Italians on the Central Coast and to the Basque populations from Bakersfield to Santa Barbara who made their own wines. Local home winemakers enjoyed their own vintages, and a number of stills were hidden in the mountains between San Luis Obispo and Bradley, California. Local wine was often stored and aged in redwood barrels that were buried in the vineyards. The Volstead Act charged the Internal Revenue Service (IRS) in the Treasury Department with enforcing Prohibition, and a Prohibition Unit was founded within the IRS. As a result, many citizens spent time in local jails after selling their wines to local customers. By 1929, there was a shift in the federal enforcement of Prohibition from the IRS to the Department of Justice, which created a Bureau of Prohibition. During these years, both Paso Robles and San Luis Obispo were known for their speakeasies.

The Klintworth and Ernst families closed their wineries during Prohibition but continued growing grapes. The York Brothers, Martinelli, Templeton and Rotta Wineries continued to grow grapes and make wine; new applications were granted by the government to the Brunetti Winery and the Pesenti Winery. The Dusi, Casteel and Dellaganna families also made wine in their vineyards during this period of San Luis Obispo County wine history.

The Central Coast dairy ranchers conspired with rumrunners to protect those who transported the liquor that was delivered by ship down the Pacific Coast to small coves in Avila, Cambria, Cayucos, Morro Bay and Spooners Cove. The dairy farmers hid the waiting trucks and drivers who transported the cargo to Los Angeles. They often used diversionary tactics, causing local law enforcement to follow them on wild chases and allowing bootleggers to escape detection. Dairy farmers and bootleggers both made money and enjoyed outfoxing the revenue agents.

The local law enforcement in San Luis Obispo County was vigilant, harsh and terrifying. People were arrested, jailed and charged large fines for making more than two hundred gallons of wine per household and for transporting and selling wine. In San Luis Obispo County, federal agents were assigned to work with local law enforcement. In 1921, local law enforcement began conducting raids and making arrests. Sylvester Dusi pleaded guilty to handling illicit liquor, as did Clifford E. Robinson, the proprietor of the Estrella Store located east of Paso Robles, in December 1921. The sentences for such crimes were levied as a choice between jail time or paying fines. These fines were used to supplement local government revenues in the cities of San Luis Obispo and Paso Robles. The typical bail for the illegal possession of alcohol was $300 in cash; the bail for illegal sale was $500.

Federal Prohibition enforcement agents arrived in Paso Robles at the end of February 1922 to conduct surprise raids on eight businesses and residences. These locations had been identified by a "government spotter," who had seen liquor purchased illegally on the premises. These spotters

Grape stomping to crush grapes has been used for thousands of years to make wine and celebrate the end of the harvest. The children seen here stomping grapes are celebrating home winemaking during the Prohibition era. *Courtesy of the Wine History Project of San Luis Obispo County.*

were often members of the local anti-saloon league, an interdenominational Protestant organization and the leading pressure group that had lobbied for Prohibition in the early twentieth century. The raids made headlines in Paso Robles on December 13, 1923, when Sheriff Ray Evans and federal officers arrested Joe Rotta. They raided his farm and found one thousand gallons of wine, although Joe claimed it was grape juice. They sealed the "juice" up for testing, and Joe was taken to jail in the city of San Luis Obispo.

Lorenzo Dusi faced charges of destroying government evidence and the unlawful sale, manufacture and possession of alcohol, according to federal Prohibition agent John H. Vail, who arrested Lorenzo on February 1, 1924, during a raid of the Dusi Ranch on York Mountain. During the raid, 1,600 gallons of wine were found in large barrels in the cellar of the ranch house. The federal and county officials sealed the barrels and left them in the cellar due to the difficulty of moving them. The agents obtained a court order to destroy the wine at the site on February 26 and found that all of the barrels were missing, except for a few large barrels that were empty. Although the evidence was missing, Lorenzo was returned to the county jail and turned over to federal authorities.

Distillery equipment, or "stills," were found everywhere in the county—from York Mountain to the Carrizo Plain. Barrels of wine were buried in vineyards between rows of Zinfandel vines. Sylvester Dusi had wine buried beneath his chicken coop, and he delivered gallons to the local hardware store when he received phone calls ordering "two hens and a rooster."

Jim Vail, the brother of John H. Vail, the Prohibition officer of the Central California District, owned the largest speakeasy in San Luis Obispo County. It was known as the Log Cabin Roadhouse and was located among cottages on Edna Valley Road in San Luis Obispo.

Amedeo Martinelli told stories of federal agents axing his barrels of sour wine while posing for a photograph. He said that after the press had gone, the agents would sit with Amedeo and enjoy his fine Zinfandel in his winery.

The enforcement of Prohibition began to ease in the early 1930s. Prohibition was clearly not producing abstinence. By 1932, wineries all throughout California had started the process of renewing their bonds to make wine again. In March 1933, President Franklin D. Roosevelt signed the Cullen-Harrison Act, which amended the Volstead Act and permitted the manufacture and sale of low-alcohol beer and wines with a limit of 3.2 percent alcohol. In most cases, the quality of the wine was very poor.

Movements favoring the repeal, which required a new amendment to the Constitution of the United States, lobbied Congress and were soon

Prohibition laws provided loopholes that allowed the use of homemade copper stills. Prohibition Bureau agents seized over seven hundred thousand stills nationwide from 1921 to 1929. *Courtesy of the Wine History Project of San Luis Obispo County.*

successful. The Twenty-First Amendment, ratified on December 5, 1933, voided the Eighteenth Amendment, which ended Prohibition but gave the right to each state to continue forbidding or revising the laws relating to the sale and production of alcohol. Some states continued their bans of alcohol; other states developed their own state controls of the alcohol and wine industries. Because of this, the laws concerning the production and sale of alcohol varied from state to state. This variation in the laws continues to be problematic for the wine industry in the United States.

Post-Prohibition (1934–1960)

Much of the California wine industry was decimated during the Prohibition era. A generation of growers and winemakers were lost, along with their winemaking skills. Many fine red and white grape varieties disappeared from vineyards. It would take three decades to restructure the wine industry, led by the founding of the Wine Institute in San Francisco and the University of California.

Most wine growers viewed grapes and wine as commodities to sell to support their families following the repeal of Prohibition. Many California winemakers—who were inexperienced—shipped inferior and tainted wines to the Chicago and New York markets in the mid-1930s, thus ruining the reputation of California wines.

San Luis Obispo County thrived during both the Prohibition era and in the years after its repeal. Local wineries were family-owned and -managed, and their grape and wine qualities were maintained. Locals planted new vineyards and opened and expanded seven major wineries. The price of local grapes increased, and many of the large California wineries were increasing their production. Some San Luis Obispo County wineries also continued to produce quality jug wine for local customers. The York Brothers and Templeton Wineries made high-quality bulk wines with premium grapes. Zinfandel grower Ignacy Paderewski focused on quality, hiring York Brothers Winery to make his wines; working together, the Zinfandel wines won gold medals at the 1934 California State Fair.

On March 16, 1936, the *Paso Robles Times* reported the largest single shipment of bulk wines ever made in one locality. The shipment of approximately seventy thousand gallons of choice Zinfandel wines had been produced at independent wineries near Paso Robles, Templeton and San Luis Obispo. Fifty thousand gallons had come from York Brothers and the Templeton Winery, which was owned by Lorenzo Nerelli. The buyer was G.C. Croce, the president of Livermore Wines Inc. (bonded after the repeal from 1933 to 1937). He supervised the shipment in specially lined railroad tank cars to the Piedmont Winery in Livermore. G.C. Croce had been a customer of A. York & Sons and York Brothers since 1901. He stated, "The Paso Robles district not only produces a better grape but remains one of the few places in the whole state where Zinfandel wines can be produced."

World War II had a major impact on the local wine industry. The area's young men enlisted and went to war, and some vineyards and wineries were abandoned by their fathers and neighbors. When the soldiers returned, old vineyards were revived, land was cleared and new acreage was planted with new grape varieties. A new era in the wine history of San Luis Obispo County transpired. The vineyards of San Luis Obispo County began producing quality grapes that were sold to wineries throughout California. The demand for these local grapes continues.

As World War II came to an end, Caterina and Sylvester Dusi purchased a second ranch in Templeton. They cleared the land and made charcoal

Above: The grape Listán Prieto (also known as the Mission grape) is believed to have originated in the Castilla–La Mancha region of Spain. *Courtesy of the University of California at Davis.*

Right: Christian Roguenant, first winemaker at Maison Deutz and Baileyana. *Courtesy of Christian Roguenant.*

Left: York Mountain Winery was owned by three generations of the York family, growing grapes and producing wine in the smallest AVA in California. *Courtesy of Neil Abbey.*

Below: The Ascension Winery label was the original name of the first wines produced by the York family in the 1890s. *Courtesy of Suzanne Redberg.*

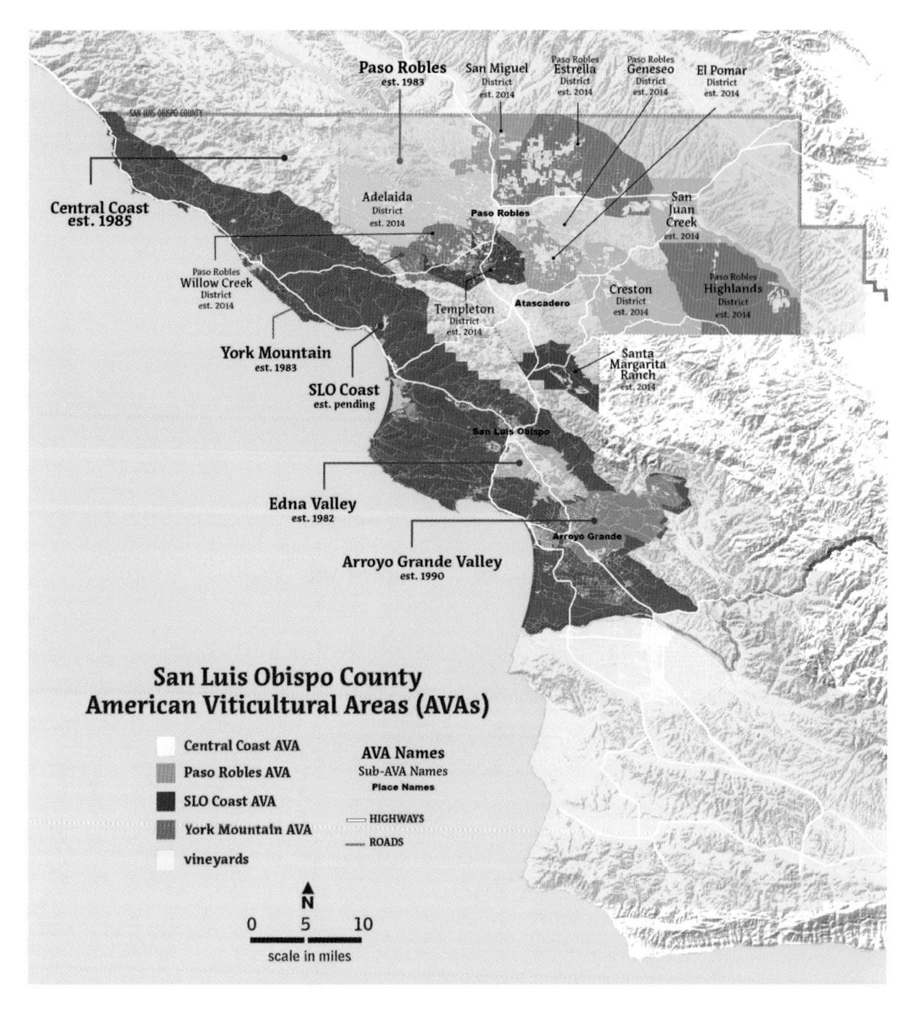

The American Viticultural Areas (AVAs) of San Luis Obispo County. *Courtesy of the Wine History Project of San Luis Obispo County, map designed by Aimee Amour-Avant.*

TIPO Chianti was the favorite wine produced by the Italian-Swiss Colony Brand, the largest table wine producer in 1937. The Italian-Swiss Colony vineyard was the second-most-visited tourist attraction in California—after Disneyland—in the 1950s. *Courtesy of the Wine History Project of San Luis Obispo County.*

Above: Aldo Nerelli and Victor Pesenti, the second generation of winemakers (1948–1969) at Pesenti Winery. *Courtesy of Frank Nerelli.*

Left: Customers brought their own containers for Romilda Rotta to fill with a ladle from the redwood barrel in the tasting room, circa 1970. Many jugs were handcrafted by local artists. *Courtesy of Aimee Armour Avant.*

Vino Sano wine bricks were blocks of dehydrated grape juice and pulp that were sold legally during Prohibition for people to make their own wine. *Courtesy of the Wine History Project of San Luis Obispo County.*

Between 1914 and 1916, Paderewski purchased 2,486 acres on two different ranches in San Luis Obispo County. He named one, Rancho San Ignacio, after his patron saint, St. Ignatius; this ranch was located on the west side of Paso Robles near Peachy Canyon Road. He named the other, Rancho St. Helena, for his wife, Helena Gorska; this ranch was located on Adelaida Road. *Courtesy of Epoch Estate Wines and photographer Chris Leschinsky.*

Top: Based on the recommendation of Tchelistcheff, HMR was the first winery to designate a local vineyard, Sauret Vineyard, as its source of grapes for the HMR award-winning Zinfandel in 1976. *Courtesy of the Wine History Project of San Luis Obispo.*

Bottom: Tom Myers in the lab at Estrella River Winery, circa 1985. Tom is known as the awesome winemakers' winemaker after forty-two harvests at Estrella and Castoro Cellars. *Courtesy of Tom Myers.*

Left: Meo Zuech, the founder of Piedra Creek, plants Gewurztraminer on a hillside in the Edna Valley, circa the early 1980s. *Courtesy of Margaret Zuech.*

Below: Chuck Ortman, also known as "Mr. Chardonnay," pioneered the technique of barrel fermentation for Chardonnay and mentored many winemakers in San Luis Obispo County. *Courtesy of Chuck Ortman.*

Left: Max Goldman purchased York Mountain Winery (formerly York Brothers Winery) in 1970. He replanted the vineyard, rebuilt the winery and made award-winning wines for the next thirty years, with his son Steve as the winemaker and his daughter Suzanne as the head of marketing and community relations. *Courtesy of Suzanne Goldman Redberg.*

Below: A nursery of grapevines at Tablas Creek. Francois and Jean-Pierre Perrin and Robert Haas imported French clones of Rhone grape varieties to supply California wineries with the highest-quality grapes from their own vineyards. *Courtesy of Jason Haas.*

Jazz trombonist and winemaker Dave Caparone is known for producing three extraordinary Italian varietals: Agliano, Nebbiolo and Sangiovese. *Courtesy of Dave Caparone.*

Castoro Cellars, founded by Niels and Bimmer Udsen, is the largest family-owned winery in San Luis Obispo County. *Courtesy of Julia Perez.*

George and Tahoma Mulder purchased their eighty-acre ranch in 1976 and built their winery, El Paso de Robles, there. Their first release was a 1981 Zinfandel. Their winery is now the site of the Castoro Cellars Tasting Room on Highway 46 West. *Courtesy of Tom Myers.*

Maison Duetz became the leader in Méthode Champenoise sparkling wines, the only winery in California and the western hemisphere to operate two French Coquard presses. *Courtesy of Laetitia Winery.*

The iconic Heart Hill Vineyard, located at Niner Wine Estates in Paso Robles. *Courtesy of Niner Wine Estates.*

A layer of morning fog covering Tablas Creek Vineyard. *Courtesy of Jason Haas.*

Spanish Springs Vineyard looking west to the Pacific Ocean in San Luis Obispo Coast wine country. *Courtesy of SLO Coast Wine Collective.*

Paso Robles wine country with the iconic oak tree growing in the vineyard. *Courtesy of Paso Robles Wine.*

Steel tanks for fermenting grape juice at Castoro Cellars Winery in San Miguel. *Courtesy of Castoro Cellars.*

A lamb resting in the biodynamically farmed Sawyer-Lindquist Vineyard in the Edna Valley. *Courtesy of the Lindquist Family.*

Hand harvesting Marsanne grapes at the Sawyer-Lindquist Vineyard in the Edna Valley. *Courtesy of Lindquist Family Wines.*

Don Othman, the inventor of the Bulldog Pup and cofounder of Kynsi Winery, with his iconic green Mack Truck. *Courtesy of the Othman family.*

Mike Sinor standing on top of his SUV, overlooking the vineyards at Bassi Ranch in Avila Valley. *Courtesy of Chris Leschinsky Photography.*

The first wine label for the jugs of Burgundy made from Zinfandel grapes that were produced at the Sylvester Dusi Winery on the Dusi Ranch in 1954. *Courtesy of Mike and Joni Dusi.*

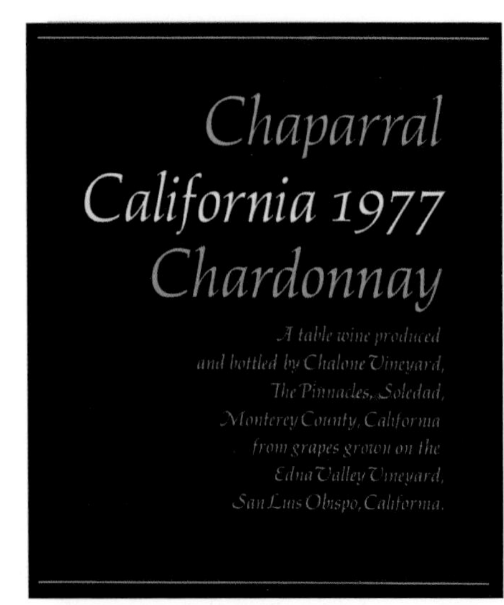

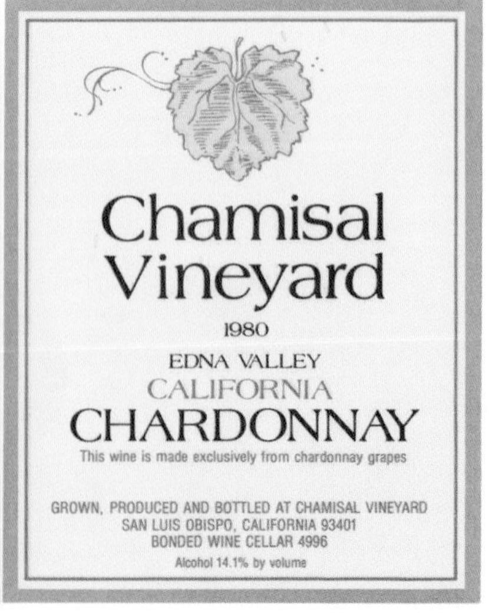

The Chaparral and Chamisal labels for Edna Valley Chardonnay, 1977 and 1980 vintages. *Courtesy of the Niven family and Chamisal Vineyards.*

with the help of Giuseppe Busi before planting vineyards of the Old-World field blends: Zinfandel, Alicante Bouschet and Carignane. This new ranch became known as the Dante Dusi Vineyards. Caterina purchased a third ranch in her own name nearby. It is now known as Caterina's Vineyard. Ten-year-old Benito was the only child at home to help work the land at the ranch, but he had already climbed up on the iconic Caterpillar Ten tractor and learned how to run it. He helped clear the land to prepare it for planting. Within a year, Dante and Guido, Benito's brothers, had returned home from the war and were working in the vineyard.

Historic Sites to Visit

Paso Robles Inn and downtown Paso Robles
Tasting room of the Glunz Winery
The iconic Madonna Inn
Thursday night farmers' market in downtown San Luis Obispo
Tour Hearst Castle and the Wine Cellar
The old Sebastian General Store and Hearst Ranch Winery wine bar
Piedras Blancas Light Station
The elephant seals at Vista Point north of San Simeon
Carrizo Plain National Monument

6

First Modern Winery Since Prohibition

HMR, Stanley Hoffman and International Awards Bring Fame to San Luis Obispo County

GRAPE VARIETY: PINOT NOIR

Introduction

The United States experienced a long postwar economic expansion, beginning after World War II and ending in the recession of 1973, which lasted until 1975. Despite this economic expansion, California wineries, which had been decimated by Prohibition, were in decline during this period. There were 414 winery licenses in 1945, and this number declined to 226 in 1969.

The 1960s started quietly with the number of California wineries in decline but with wine production gradually increasing. Consolidation in the wine industry continued, with names like Gallo and United Vintners growing larger. Mediocre wine was produced in many small family-run wineries, as fortified wines were far more popular than table wines. Unlike today, only 10 to 15 percent of California's wineries at the time produced a product that would be identified as fine wine. By the end of the 1960s, five trends were occurring and surprisingly moving in opposing directions: the consumption of table wines increased dramatically; large wineries began to lose control of their market share; talented people left their jobs to become winemakers, establishing small wineries and planting classic grape varieties; wine shops opened in Los Angeles, San Francisco, Sacramento and Berkeley; and large commercial vineyards were planted with irrigation, replacing dry-farming.

One cardiologist from Southern California changed the world of wine in San Luis Obispo County in the 1960s when he decided to plant grape varieties from Burgundy and Bordeaux. He hired renowned wine consultant Andre Tchelistcheff to advise him on winemaking and the construction of the first modern winery in San Luis Obispo County. His wines won international awards and focused worldwide attention on San Luis Obispo County.

In the 1960s, the San Luis Obispo County wine trade continued as it had in the past, dominated by small family wineries. These were then run by the children and grandchildren of their founders, including the Yorks, the Nerellis, the Rottas, the Pesentis, the Dusis, the Martinellis and the Casteels. Grape growers continued to dry-farm at relatively small vineyards. Gradually, more acres were planted by local growers, including Richard Sauret and Mel Casteel. Local grapes continued to be sold to producers throughout the state.

The agricultural advisor to the county, Jack Foote, witnessing the decline of the almond industry, began planting experimental vineyards with new grape varieties. He advised local farmers to plant vineyards on their fields of grain and cattle ranges. In the late 1960s, the first large irrigated vineyard was planted by grain farmer Robert Clark Young with Cabernet Sauvignon vines for producers in Napa. Dry-farming had dominated grape growing in the county for the past 160 years, but Young decided that irrigation would increase his vineyard's yields after discussing dry-farming techniques with Frank Pesenti. This choice of using irrigation for growing grapes initiated a new era that would eventually strain the water resources of the county.

The small family wineries, which were then managed by the second, third and fourth generations of the county's winemaking families, continued producing premium grapes and wine.

Rotta Winery: The Second Generation, Mervin Rotta (1924–1997); and Third Generation, Robert Giubbini (1941–Present) and Michael Giubbini (1953–Present)

Three generations of the Rotta family grew grapes and made wine on their farm. Mervin Rotta, the son of Clement and Romilda, worked in

Mike Giubbini, the grandson of Clement and Romilda Rotta, was small enough to squeeze inside the tanks and clean them. Here, he is seen "taking a breather," circa the late 1950s. *Courtesy of Mike Giubbini.*

Romilda Rotta greeted customers at any hour to sell Rotta wine. The winery made at least four types of wine, including a Zinfandel Rosé. The winery's oldest Zinfandel wine was aged seventeen years in redwood tanks, circa the 1960s. *Courtesy of Mike Giubbini.*

the vineyards while transporting crops in his trucking business. Irene, his sister, lived in the Bay Area with her husband and sons, Robert and Michael Giubbini. Bob and Mike worked during school vacations in the 1960s, learning the art of grape growing. As the younger child, Mike was small enough to crawl into the redwood tanks to clean them. He helped make the wine and bottled it in glass jugs, applying the famous Rotta Label. Mike loved working with his uncle and grandparents; although he first pursued a career with the California Department of Forestry and Fire Protection (CAL FIRE), wine was his passion. Mike also worked with the Martinelli family in their vineyards on Ridge Road. Mike fondly remembers Rina Martinelli serving him red wine at lunch; he pointed out that "he often found an earwig floating in his glass."

In the 1960s, Rotta Winery's reputation spread to surfers and hippies on the Southern California coast and north to Big Sur. Wine lovers drove hundreds of miles to fill their jugs in the new tasting room, which had been built from a large redwood cask. Local California Polytechnic State University (Cal Poly) students would drive into Romilda's driveway and trip the bell under her mat; she then would bring jugs of wine right to their car windows. In 1973, the *Sunset Book on California Wine* described the three local wineries, York Mountain, Pesenti and Rotta, stating, "Much of the future of wine in Southern California lies with the new acreage developing around Templeton."

Clement died in 1963. Mervin and his wife, Jean, moved to the property in 1967 to maintain the vineyards and the winery with Romilda. Mike lived with his grandmother while attending Cal Poly and helped her farm until her death in 1976. In the same year, Rotta Winery and Vineyards was sold to new owners, who changed its name and label to Las Tablas Winery.

Mike Guibbini continued to dream of establishing the Rotta Winery once again. When the new owners sued his grandmother's estate, years of litigation ensued. Mike fought to regain control of the winery; he was able to acquire forty acres of the Rotta Winery in 1990. Steve Pesenti joined him as a partner to expand the winery and vineyards. Mike hired experts to prepare a historic structure report to provide documentary, graphic and physical information on the property and its existing condition in order to preserve and maintain its original buildings. The winery was modernized and produced several varietals, including the traditional Zinfandel.

On December 22, 2003, the San Simeon earthquake struck with a magnitude of 6.6, causing over $250 million worth of damage to local property, including Rotta Winery. Mike and Steve continued operations and repaired the damage to the new winery. The second economic blow to all of the winery owners in North County occurred during the 2008 financial crisis. Rotta Winery was sold to new owners in 2013 for an undisclosed sum. Mike is currently retired and lives with his wife in Templeton, surrounded by grapevines. Steve Pesenti has since retired in Idaho.

Pesenti Winery: Third Generation, Steve Pesenti (1949–Present) and Frank Nerelli (1949–Present)

When third-generation cousins Frank Nerelli and Steve Pesenti replaced their parents at Pesenti Winery in the 1990s, their focus shifted to producing high-quality premium wines. Twenty-one-year-old Frank became the winemaker in 1970. He developed new techniques, anticipating the changes that were coming with the California "wine revolution." Frank produced new varietals: Cabernet Sauvignon, Muscat, White Riesling, Sauvignon

Blanc, Grey Riesling and Gewurztraminer. Zinfandel continued to be the most famous varietal. The famous wine writer Gerald Asher wrote in an article for *Gourmet Magazine* in 1994 that the Pesenti three-liter jug of Zinfandel was "the best value in red wine anywhere in the world."

Pesenti Wine continued to have a loyal following among shepherds throughout central California. The Basque populations from Bakersfield to Salinas bought their wines in barrels and bota bags. The tradition continued as sheepherding continued to thrive in North County during the 1970s; each sheepherder received a daily ration of bread and cheese and Pesenti wine.

Pesenti Vineyards and Winery was sold to winemaker Larry Turley in 2001. Frank opened his own winery, ZinAlley, and Steve Pesenti joined Mike Giubbini as a partner in the new Rotta Winery. Turley Wine Cellars remains famous for its Zinfandels, among other varietals. Larry Turley values the wine history of the Pesenti Winery, celebrating it with photographs, films and Zinfandel exhibits at the Turley Wine Cellars Tasting Room. Larry Turley has also acquired the historic vineyards of Amedeo and Bruno Martinelli and the Ueberroth family; he has restored the historic structures and vineyards. Larry Turley and his family continue to work to preserve historic vineyards with old vines dating back to the 1800s in California. Turley makes forty-seven separate wines from over fifty vineyards supervised by the director of winemaking, Tegan Passalaqua, who is a founding member of the Historic Vineyard Society.

Frank Nerelli (1949–Present): The Third Generation Establishes ZinAlley Winery

Frank Nerelli, born on April 7, 1949, grew up in the Nerelli family home just above the Pesenti vineyards. Frank began working at the winery at the age of ten; he harvested grapes in redwood lug boxes, pruned vines, stuck labels on jugs and applied sealant to wine bottles. Frank also farmed grain and other crops on the Pesenti family farm; he was the only one of Frank Pesenti's twenty grandchildren who continued to work at the Pesenti Winery until it was sold to Larry Turley in 2001.

There is a strong tradition among Italian sons and daughters to join their fathers and grandfathers in the family businesses. Frank served in the armed forces after high school and returned home to work with his father, Aldo, his uncle Vic and his grandfather Frank in the vineyards and on the farm. Pesenti Winery was famous for its jug wine throughout California.

In the 1970s, as the demand for higher-quality wine increased and Paso Robles became known as "a grape-growing area with potential," Frank Nerelli pondered his future in local winemaking. He wanted to select the right variety to make in his own style. His passion was Zinfandel, the "heritage grape" of the county. He knew that Zinfandel presented a real challenge for both growers and winemakers. To quote Frank many years later, "It is worth it if you like a challenge." When Frank became the winemaker at Pesenti Winery, his father and uncle wanted Frank to continue in their traditional winemaking style, rather than experiment with new varieties.

LAND TENDS TO BE PASSED DOWN THROUGH THE GENERATIONS IN ITALIAN FAMILIES

Frank had purchased his own land from his uncle Vic Pesenti in 1972; Vic had purchased the same parcel from his father in 1947. Frank built his home and planted barley and walnuts on the land immediately. He spent the next four years preparing the soil and doing research on the best Zinfandel clone to produce premium grapes in the Templeton Gap. He found that clone in Amador County. Frank replaced his crops with dry-farmed, hedge-pruned Zinfandel vines that he had grown in his own style to produce rich and very complex Zinfandel wines. Frank's vineyard style was in vogue during the first half of the twentieth century, but it is rarely seen today. The vine was the most important part of his meticulous vineyard management. He carefully observed each plant and made adjustments in his methods when necessary. Today, he still works in the vineyard alone, except at harvest time.

Frank worked at Pesenti Winery as winemaker for more than three decades. The sale of the winery to Larry Turley in 2001 gave him the financial freedom to pursue his own dreams. Frank established his own winery, ZinAlley Winery, elevating the winemaking traditions of the Nerelli and Pesenti families in his small family-owned winery. He is well known for his unusual, elegant Zinfandels and Ports. Frank can usually be found in the tasting room or in his vineyard right outside the door.

Bill York (1904–1984): Third Generation at York Mountain

Around 1929, Wilfrid (Bill) York left York Mountain and the winery to attend UC Berkeley to major in science and engineering. He loved the San Francisco Bay Area and enjoyed the arts. He had a fine singing voice and played both the violin and the piano. He painted landscapes and seascapes in oil throughout his life. After graduating from college, he married Dorothy May Osborne and found employment in San Francisco at Wells Fargo Bank. He also taught music at the San Francisco Conservancy and played the violin in the Wells Fargo Symphony Orchestra. Bill loved the stars and night sky; he built his own telescope and ground the lens with his own hands. He was one of five children and hadn't planned to be the next winemaker at the family winery. But because none of his siblings expressed an interest in taking on the vineyards or winemaking, Walter and Lillian York, Bill's parents, summoned him back to York Mountain. So, Bill and Dorothy, along with their baby daughter, Joan, returned to the Templeton Gap in 1936.

In 1944, ownership of the York Brothers Winery was transferred to both Walter's son Wilfrid (Bill) and Silas's son Howard. Dorothy gave birth to the couple's second daughter, Jan York, in 1945. Howard decided to pursue a career in engineering in 1954 and sold his ownership to Bill.

Nathan Chroman, a wine writer for the *Los Angeles Times* and author of *The Treasury of American Wines*, described the production of Zinfandel at York Brothers Winery in the 1940s, saying "[It was] produced in a style very different from that of northern and Southern California—the wines were of bigness, complexity and good varietal character." These wines were often purchased in bulk in barrels.

Bill mapped out his own plan for an expansion of wine production. Bill bought additional grapes from local growers, including the Busi, Dusi, Ernst, Nerelli and Venturini families. Bill's grapes were harvested with the help of his friends, neighbors and family members. By 1953, Bill York was producing eighty thousand gallons of wine, including Zinfandel, Claret and Burgundy. He won five state fair awards in the 1950s.

The year 1960 was a turning point in the history of York Brothers Winery. That year, it was awarded gold medals for its Zinfandel at the state fair in Sacramento and the Los Angeles County Fair. These achievements were followed by a series of events that destroyed crops and wine production at York Brothers Winery. In 1962, an invasion of deer destroyed the vineyards. In 1964, a tornado hit York Mountain and ripped the roof off the winery,

Bill York on a tractor in a York Mountain vineyard. *Courtesy of Jan York.*

damaging one fermentation tank and destroying the other. In 1968, a freak snowstorm on York Mountain damaged the Zinfandel grapes.

In 1970, Bill York decided to retire and sell York Brothers Winery, which had been established in 1882 and owned by three generations of York winemakers. He sold the property to legendary enologist Max Goldman, who had retired to Malibu, California, with his wife, Barbara, after a long and successful career in the wine industry. Both men and their wives were guests at a dinner party in the beach town of Cayucos. During the course of the dinner, Bill mentioned that the winery was for sale. He described its history, including the fact that Paderewski had chosen York Winery to press his grapes and make his famous award-winning Zinfandel. Max was overcome with emotion. He had been an accomplished pianist since childhood, and the works of Paderewski were among his favorite compositions. It seemed as if the fates had conspired to bring these two winemakers together to pass the York baton to Max Goldman. Max and his children, Steve Goldman and Suzanne Redberg, replanted the vineyards with new varieties, restored the winery and family home, added a legendary tasting room and produced award-winning wines for the next three decades, beginning in 1970. Max renamed the winery York Mountain Winery. Bill York died on June 9, 1984, leaving a rich legacy for San Luis Obispo County.

Martinelli Vineyards Preserving Amedeo's Legacy: Bruno Martinelli (1949–Present)

Bruno Martinelli was born in 1949 and lived in an orphanage in Bologna, Italy, for most of his childhood. In the aftermath of World War II, his family members were separated from one another because of economic hardship. He eventually joined his family in Posina, Italy, after his mother, Rina, brought all five of her children home to live together. Bruno was adopted in Italy in 1959 by Amedeo Martinelli, his mother's new husband.

Bruno, like many other Italian children, had endured many hardships following World War II, including the failing economy, hunger and homelessness. So, when he arrived in Templeton on Christmas Eve in 1960 from Posina, Italy, he faced enormous challenges. He spoke no English and knew nothing about farming or the American culture. He attended the Templeton schools and quickly learned English. He enjoyed athletics and became a great fan of volleyball and basketball. Coaching became his avocation, and he coached volleyball teams in Paso Robles for more than forty-eight years.

Bruno became an American citizen in 1965, mindful of his dedication to Amedeo's legacy. He graduated from Templeton High School and joined the navy. He met his sweetheart, Debbie Martin, in Atascadero while he was home on leave, and he proposed within two weeks. They were married a year later, in 1972, and raised two children, a son, George, and a daughter, Deanna. After leaving the navy, Bruno worked with the Templeton School District, maintaining the buildings and grounds. Later on, he worked in the San Ardo oil fields to support his growing family. He also worked in the Martinelli vineyards in his free time.

Bruno took great pride in the vineyards and continued to help his mother maintain them. Rina sold her first growth of Zinfandel to Pesenti Winery but saved the second growth to make wine for her family. In 1986, when Rina was no longer physically able to manage her property, Bruno and his family purchased a mobile home and moved to the Ridge Road property in Templeton to help his siblings care for her and maintain the vineyard. They continued to sell their grapes primarily to Pesenti, but occasionally, they sold to the new wineries that were opening in Paso Robles during the 1980s, including Arciero Brothers.

Rina required a quadruple bypass on September 30, 1991, and she lived until December 28, 1998. At the time of her death, her three youngest children, Maria, Bruno and Silvana, inherited the property. Maria bought her siblings' shares of the property and became the sole owner of the vineyard.

Bruno and his family moved to Paso Robles in June 1999. He described the Martinelli vineyard as his pride and joy. His cheerful personality and his wonderful storytelling continue to be enjoyed by all who know him. He continues to coach volleyball teams throughout San Luis Obispo County. Bruno and Debbie have since retired, and they purchased a new home on Creston Road. Bruno enjoys growing pomegranates.

Maria sold the property in 2014 to Larry Turley, the owner of Turley Cellars. Turley maintains this historic vineyard and has restored the Martinelli buildings on the property. Bruno enjoys visiting the tasting room to see the old film of Amedeo and his family celebrating with Zinfandel wine.

TWO LEGENDARY VINEYARDS: DANTE DUSI (1925–2014) AND BENITO DUSI (1933–2019)

Dante and Benito were devoted to their vineyards and their Zinfandel vines. All three brothers worked together and shared meals, laughter, politics, friendship and family celebrations their entire lives. Guido spent hours working with Benito in the vineyards and working on the fragile pump known as the "Buddha." Their parents, Sylvester and Caterina, were hardworking entrepreneurs; they expected no less from their children and grandchildren.

Benito's Zinfandel vineyard was planted with the original old vines, which dated back to the mid-1920s. Benito lived on the ranch with his parents, and his mother, Caterina, became his business partner after Sylvester's death in 1964. This partnership negotiated a contract to sell 95 percent of the

Benito Dusi Zinfandel harvest annually with Dave Bennion, the founder of Ridge Vineyards, over a Sunday lunch of polenta and stew. Dave Bennion had a passion for Zinfandel and was driving the California highways trying to find old vine, dry-farmed vineyards from which to source grapes for his new winery. He spotted the Dusi Vineyards while driving, pulled into the driveway and knocked on the front door. Ridge and the Dusis have continued this relationship for over fifty years.

Benito saved the second growth on the vines for home winemakers, allowing them to harvest their own grapes or send them to John Daume at the Home Winemaking Shop in Woodland Hills for amateur winemakers to purchase. Benito remained a bachelor but loved his nieces and nephews and their spouses and children as his own. Benito lived in his parents' house, and the family continued to work with Benito in his vineyards until his death. He was particularly close to his nephew Mike's wife, Joni Dusi, who cared for him until his death. His family continues to farm the iconic vineyard and sell grapes to Ridge Vineyard.

Dante Dusi married Dorothy Steppie, and they raised three children in the Dante Dusi Vineyard, which had been purchased in 1945 by Sylvester and Caterina Dusi. Dante cleared the land with his family and Giuseppe Busi when he returned from service in the U.S. Coast Guard. He planted the vineyard with his family in stages and farmed it on the weekends until he retired from Madonna Construction in 1977, at which time, he became fully devoted to his vines. The vineyard was planted with the same field blend as the Benito Dusi Vineyard. Small amounts of Carignane, Mission, Alicante Bouschet, Petite Sirah and Grenache were planted. In contrast to Benito Dusi, Dante always sold his grapes to a variety of winemakers. In the 1960s, the reputation of Dusi grapes had spread far beyond San Luis Obispo County lines. Paul Masson and Monterey Wines were the first large wineries to purchase Dante Dusi grapes. Many local winemakers have since been customers, including the York family, Dave Caparone, Tobin James, Robert Nadeau, Doug Beckett, Steve Dooley and Turley Cellars Winery, just to name a few.

After Dante's retirement, his son Mike's wife, Joni Dusi, worked with him in the vineyards, and his granddaughter Janell was his constant sidekick. Dante mentored many growers and winemakers over the years, including Janell, who has become a well-known winemaker in her own right. All Dusi vineyards are maintained and harvested by Dusi family members. Dante's son Rick moved from San Diego in 1996 to join his father full time in the vineyards. Mike joined them in 2000. Mike has purchased more vineyards and planted new varieties over the last ten years. All three of Mike's children, Michael, Matt and Janell, continue to work in the family business. Janell continues to honor Dante's legacy with the "Dusi blue" labels on her wine bottles and family photographs and exhibits in her tasting room.

The First Tasting Room on United States Highway 101

In the 1950s, as grape prices were falling, Sylvester had a surplus of his own grapes at harvest time, so he decided to make and bottle Dusi wine. Benito purchased the redwood tanks and equipment he needed in San Francisco. Sylvester received the certificate of label approval under the Federal Alcohol Administration Act on July 15, 1954, for the Sylvester Dusi Winery. The label was for California Burgundy. He built a winery and opened his own tasting room on the ranch. His son Benito became the winery's winemaker and applied for the brand name Dusi Wine California Zinfandel on July 5, 1955. Benito chose the new label from a catalogue.

The tasting room was the first of its kind on U.S. Highway 101, the major scenic highway in California between San Diego and San Francisco. The tasting room welcomed visitors from 1954 to 1967, including the military members who were serving at Camp Roberts. Surprisingly, many of the visitors in the tasting room were Europeans.

The Dusi family abandoned commercial winemaking in the mid-1960s to focus on premium grape production, which the third, fourth and fifth generations of the family still do today. Janell Dusi continues to make premium wines labeled with family legends.

Jack Foote (1923–2005): New Varieties Grown in Experimental Vineyards

San Luis Obispo County has an agricultural advisor funded by the University of California Statewide Extension Department of Agriculture. The advisor's long tradition of providing research and consultation to local farmers dates back to 1889, when the first agricultural advisor planted over 150 grape varieties east of Paso Robles.

Jack Foote, an agricultural advisor who was assigned to the county in the 1960s, was researching new crops to sustain the local economy. He planted many grape varieties in experimental vineyards located in various regions of the county. He was able to prove that a number of new varieties could be grown successfully in the county. The success, he determined, was dependent on the marine influence, the soils and the microclimate. Jack shared his research on grape growing with local farmers.

In 1964, Jack Foote advised Stanley Hoffman to plant ten acres of Pinot Noir grapes on his ranch. He was also the advisor to those pioneers who planted Chardonnay grapes in the Edna Valley in the 1970s and to Cliff Giacobini, the founder of Estrella River Winery. Jack convinced Cliff that growing grapes might be more profitable than cattle ranching. Jack Foote had a profound influence on local wine history.

A photograph of Andre Tchelistcheff. Stanley Hoffman was the first winery owner to hire the famous Andre Tchelistcheff as a wine consultant. Andre trained winemaker Michael Hoffman to make world-class wine. *Courtesy of the University of California at Davis, Special Collections.*

Andre Tchelistcheff (1901–1994): Godfather of California Winemaking

Andre was given the moniker "godfather of modern California winemaking" and was the most influential winemaker in California since the repeal of Prohibition in 1933. He was also known as the "Maestro" for his impressive winemaking. He pioneered the modern techniques that changed California winemaking when he moved to Napa in 1938 to work for Georges de Latour, the owner and founder of Beaulieu Vineyard. Andre introduced new winemaking techniques, such as the cold fermentation of white and rosé wines, the control of malolactic fermentation in red wines and the use of small oak barrels for aging, all of which profoundly changed the way California wines were made.

Andre Tchelistcheff trained many California winemakers and pioneered the study of viticulture and terroir in Napa Valley. His own wines produced at Beaulieu Vineyard became the benchmark standard for California winemakers. After Andre retired in 1973, he consulted with wineries and winemakers on the Central Coast, including the Hoffman Mountain Ranch Winery (HMR) in San Luis Obispo County. Hoffman hired Andre to work with HMR starting with the 1973 harvest. Andre trained Stanley's son, winemaker Michael Hoffman. Andre was awarded the James Beard Award for Outstanding Wine, Beer and Spirit Professionals. He died of lung cancer in 1994.

First Modern Winery Since Prohibition: HMR, Stanley Hoffman (1920–2017), Michael Hoffman (1954–Present) and David Hoffman (1949–Present)

Dr. Stanley Hoffman established the first modern large-scale winery in San Luis Obispo County with the expertise of wine consultant Andre Tchelistcheff in 1972. Stanley is considered the "godfather of winemaking" on the Central Coast. His legend has inspired future generations of winemakers in San Luis Obispo County to make world-class wines. The vineyard he planted in 1964 is one of the oldest Pinot Noir vineyards in California, and that vineyard brought him the recognition of being the first to grow Pinot Noir grapes in San Luis Obispo County. Stanley was also known as the first to plant French Burgundian and Bordeaux varieties in the county after Prohibition.

Stanley grew up on a farm in Terre Haute, Indiana. He helped his dad with farming and home winemaking during Prohibition. He pursued an education in medicine, specializing in cardiology. He moved to Southern California to open his own practice at the Los Angeles International Airport in 1950. Terry, his high school sweetheart, and Stanley raised five children in Beverly Hills and on their ranch in Thousand Oaks, California. Their sons, David and Michael, are part of the winery's story.

Stanley and Terry studied and tasted fine wines, traveled to European vineyards and studied the soil and varietals. It is not known when Stanley may first have dreamed of planting his own vineyard or making his own wine; however, when Stanley's friend Ernie Fenders located a beautiful 1,200-acre ranch property planted with almond groves in the hills west of Paso Robles, Stanley was extremely interested in buying. In 1961, Stanley

Stanley Hoffman, the "Godfather of the Central Coast," pioneered modern winemaking, circa 1975. He built the first modern winery in 1972 with a plate-and-frame filter press and steel tanks. *Courtesy of Michael Hoffman.*

was able to trade his ranch in Thousand Oaks for the ranch along Adelaida Road, an area that had been well known during the previous one hundred years for running cattle and growing grains. The Hoffmans named the property Hoffman Mountain Ranch, and the family went there during the weekends and summers.

Stanley sought the advice of local agricultural advisor Jack Foote to determine what crops he should plant on the ranch. Stanley and his family had the original almond orchards growing as a first crop and then added walnuts. Jack Foote was extremely helpful and counseled Stanley on the best practices for his production. Stanley sold his crops locally. On subsequent visits, Jack recommended that Stanley plant a vineyard with Pinot Noir grapes. Stanley enjoyed Pinot Noir wine—it was his personal favorite—so he hired vineyard manager John Whitener to plant and manage his vineyard. The elevation of the vineyard where Stanley planted ten acres of Pinot Noir on its own roots in 1964 was 1,725 feet above sea level. Cool sea breezes from the Pacific Ocean caressed the vines in this location. Jack Foote also advised Stanley to plant Cabernet Sauvignon, Chardonnay and Franken Riesling for a total of two red and two white grape varieties in his vineyard.

Stanley Hoffman is recognized as a legend for introducing modern wine technology to San Luis Obispo County and the Central Coast. He was the first in the county to hire world-famous Andre Tchelistcheff as his wine consultant. Hoffman Mountain Ranch was the first modern commercial winery to be built in the county after Prohibition. It was designed by David Hoffman and was constructed between 1972 and 1975. With the assistance of Tchelistcheff, the winery purchased the first newly designed stainless-steel fermenting tanks and cutting-edge equipment, along with new French oak barrels. HMR wines were made using the first modern winemaking equipment ever used in the county.

Andre Tchelistcheff taught winemaker Michael Hoffman to make world-class wines. The Hoffmans were the first in the county to win an international wine competition. HMR won gold and double-gold medals for HMR Chardonnay in the International Wine and Spirit Competition in London in 1979. This award focused worldwide attention on the wines produced in Paso Robles and San Luis Obispo County. HMR was featured by wine writers in national magazines and international publications, which, at the time, was rare publicity for the Central Coast. Stanley was the first local winery owner to be featured on the cover of *Wine and Vines Magazine* in 1978. HMR was one of the first twentieth-century winemakers to sell San Luis Obispo County wines in Asia and Europe. Public recognition of the quality of HMR wines established the Adelaida District as a grape-growing environment. HMR sourced Zinfandel grapes from Richard Sauret. Andre Tchelistcheff suggested HMR designate the source of its grapes, the Sauret Vineyard, on the wine label of its award-winning 1976 Zinfandel. Although

This aerial photograph of the Hoffman Mountain Ranch shows the steep terrain where Stanley Hoffman planted the first Pinot Noir in San Luis Obispo County in 1964. *Courtesy of Michael Hoffman.*

the vineyard designation is often shown on wine labels today, HMR was the first to do so in San Luis Obispo County.

The Hoffman family was very outgoing and supported music events and local charities. They were the first to host the annual San Luis Obispo Mozart Festival (now known as Festival Mozaic) at a winery. Their winery was also the first to raise funds to support the festival. The tradition of fundraising for nonprofits by wineries was born and continues to be a major source of fundraising for local charities today.

Although economic pressures and dissent among investors forced the Hoffman family to lose control of HMR in 1982, Stanley continued his medical practice as a cardiologist in Paso Robles, and he was beloved by all who knew him. Stanley is still revered by local winemakers, including Marc Goldberg and Maggie Da' Ambrosia, the owners of Windward Vineyard in Paso Robles. Marc and Maggie specialize in estate-grown, Burgundian-style handcrafted Pinot Noir. Stanley's friendship was treasured by Marc, and HMR wines are still in his cellar, a remarkable tribute to the Hoffman legacy and legend.

Historic Sites to Visit

Adelaida Vineyards
Le Cuvier Winery
Windward Vineyard
Downtown Cambria and Moonstone Beach
DAOU Vineyard and Winery

7
The California Wine Revolution Invades San Luis Obispo County

Grape Variety: Chardonnay

Introduction

At the start of 1970, San Luis Obispo County had three successful wineries with estate vineyards, York Mountain, Rotta and Pesenti, all located in Templeton. The long tradition of over one hundred years of winemaking in San Luis Obispo County continued without interruption. The local Zinfandel growers included Dante, Caterina and Benito Dusi, Richard Sauret, Rini and Bruno Martinelli and Mel Casteel with vineyards dating back to the early twentieth century. By 1972, there were approximately nine hundred acres of grapes planted to the east and west of Paso Robles; these growers were selling to winemakers in Los Angeles, Monterey, Napa and Sonoma Counties.

Robert Clark Young (1916–2011), a 1936 Olympic athlete and grain farmer in Shandon, established the first large commercial vineyard in San Luis Obispo County since Prohibition. He laid out his vineyards (now known as Rancho Dos Amigos) in 1963 with irrigation, planting Zinfandel and Carignan. Young was advised by county agricultural advisor Jack Foote to add more red varieties, such as Cabernet Sauvignon. Changes in tax laws encouraged investments in agriculture; Herman Schwartz purchased 2,500 acres in 1969 and organized a partnership of Hollywood personalities to invest in 500 acres of vineyards near Whitley Gardens. The vineyard was planted with Zinfandel, Cabernet Sauvignon and Merlot grapes in 1973. Most of the grapes grown in both vineyards were shipped north to Napa and Sonoma Counties.

The California wine revolution created a demand for many new grape varieties in the 1970s, especially with the rise in independent winemakers learning the craft from home winemaking classes or majoring in enology and viticulture at UC Davis and Fresno State. These new winemakers abandoned their former careers and moved to San Luis Obispo County to plant vineyards that were typically less than forty acres in size. These growers gradually added wineries but struggled with marketing their wines, and most San Luis Obispo County wines were unknown to restaurants and wine shops in the 1970s and 1980s.

The first large modern winery with a scientific laboratory and state-of-the-art winemaking equipment was built in Paso Robles and staffed with UC Davis graduates.

The Edna Valley, once the home for Mission grape vineyards, was planted with fields of grain. Foote planted experimental vineyards throughout the county in the 1960s and 1970s to advise farmers on new crops to replace the declining almond industry. His research, funded by the University of California, encouraged growers to plant new varieties of grapes on farmland and cattle ranges.

North and South San Luis Obispo County: The History of the Great Divide

San Luis Obispo County is known for its diversity in geography, geology, climate and weather. The Cuesta Pass, a summit rising to an elevation of 1,549 feet, separates the north county and south county. The rugged terrain through the pass was an obstacle to trade, commerce and local travel until the railroad connected the agricultural center of north county, Paso Robles, with the City of San Luis Obispo in south county in 1894.

San Luis Obispo was the county seat and the gateway to Port Harford in the town of Avila on the coast in south county. Goods, building supplies, mail and travelers from Los Angeles to San Francisco had access to the county primarily by ocean, landing at Port Harford. The city of San

Luis Obispo supported banks, real estate development, grand hotels, restaurants and a variety of local businesses. California Polytechnic School was established nearby in 1901, and it was renamed California Polytechnic State University in 1972.

The economies of the county, both north and south, were also shaped by geography and climate. Raising cattle and sheep; mining for chromium, mercury, manganese, gold and uranium; and planting orchards, barley, oats and wheat originally drove the local economy in north county. Gradually, grapes and wines have replaced many of these commodities, and this shift started in the 1870s. The dairy industry's butter and cheese, the fishing industries, mineral exports and agriculture were all historically important in south county.

North County: Estrella River Winery, the Launching Pad for Wine Legends

Gary Eberle

This story starts with a family celebration in Florida for the seventieth birthday of Laura Eberle, the mother of half-brothers Gary Eberle and Cliff Giacobine. There, the brothers swapped stories. Cliff was tired of the pace of life in Los Angeles and wanted to move to the Central Coast to purchase a cattle ranch. Gary had graduated from Pennsylvania State University with a degree in biology and entered a doctoral program at Louisiana State University (LSU) to study cellular genetics as a national science fellow. LSU professor Harold Berg and Gary shared a passion for opera and enjoyed fine French Cabernets and Bordeaux from Berg's wine collection. Gary started reading about wines, and although he had never visited a winery or met a winemaker, he knew he wanted to join the California wine revolution. His first step was to join the enology and viticulture programs at UC Davis.

Both men made momentous life decisions in 1971. That year, Cliff consulted with county agriculture advisor Jack Foote to discuss cattle ranching. Foote advised Cliff to plant a vineyard in San Luis Obispo County. Instead, Cliff purchased land on the Estrella Plain east of Paso Robles and learned the cattle business. Meanwhile, Gary applied to UC Davis and was admitted to the doctoral program in enology.

Gary Eberle planted the first commercial vineyard in Syrah with its own roots in 1975. He was the first California winemaker to make a varietal with 100-percent-Syrah grapes. He established his own Eberle Winery in 1979. *Courtesy of Gary Eberle.*

The relationships Gary developed with his professors were crucial to his success as the cofounder of two of the most influential wineries in Paso Robles. During his studies, Gary traveled to San Luis Obispo County with viticulturist Dr. Harold Olmo to take soil samples. Gary's research and analysis resulted in his professors advising him to plant grapes in the Estrella River Valley region, the same area where the padres of the San Miguel Mission had planted and cultivated their vineyards.

The Eberle and Giacobini families eventually became partners. Between 1973 and 1982, Gary and Cliff planted a large vineyard of seven hundred acres with twelve grape varieties and installed irrigation. They also designed and built the largest modern winery in the county. They named it Estrella River Winery, honoring the terroir and capturing the "sense of place." Estrella River Winery became the launching pad for the new generation of legendary winemakers in North County.

North County Legends of the 1970s

John Munch

John Munch, the co-owner of Le Cuvier Winery. He sourced his fruit from Paso Robles dry-farmed vineyards and let the grapes do the work. His non-interventionist approach is legendary. *Courtesy of Julia Perez.*

John and his late wife, Andree Guyon, moved to Paso Robles in 1978. John, a man of many talents, including cabinetmaking and restoring Victorian houses, bought property on Vine Hills Lane in the Adelaida Hills, where he built his home and established his winery, Le Cuvier. John became a winemaker by accident. He was asked to research the possibility of producing sparkling wine by a group of French investors. At the time, there was a growing interest in making Champagne in the United States. John learned the traditional methods and made sparkling wines with grapes sourced from Estrella River Vineyard and vines grown in Shandon. The wine was successful, but the partnership with the French investors failed.

John established his first winery, Adelaida Cellars, with his 1981 vintage. To the surprise of Gary Eberle and Tom Myers, the winemakers at Estrella River Winery at the time, John won awards with his wine. John and his colleagues Niels Udsen and Bill Sheffer continued to buy bulk wine in barrels to blend for their own labels until Estrella River Winery filed for bankruptcy in 1987.

John moved to new locations to make wine under both of his labels, Adelaida Cellars and Le Cuvier. His winemaking style is one of non-intervention or allowing the grapes to do the work. His extensive wine library holds local wines spanning five decades. His recollections of wine history in San Luis Obispo County are valuable, entertaining and insightful.

Tom Myers

Tom graduated with a master's degree in enology from UC Davis in 1978. He focused his studies on the science of winemaking.

Tom joined the staff of the legendary Estrella River Winery in Paso Robles as an assistant winemaker in 1978. Estrella River Winery was the first to establish a lab for testing its wine with state-of-the-art equipment and technology. The wines Tom made there have served as the prototypes for the trends and styles in the Paso Robles AVA, including Estrella River Cabernet Sauvignon, Castoro Cellars Zinfandel and red blends. Tom is recognized as one of the great Zinfandel winemakers. He said, "The distinctive and amiable characters of Zinfandel entitle it to rank among the noble varieties of the world. It is our heritage grape." He became the head winemaker at Estrella River Winery in 1982.

Tom joined his friends Niels and Bimmer Udsen as a winemaker for Castoro Cellars in 1990, making wines for the brand and for its many custom crush clients for more than three decades. Tom sees wine as one of the great foundations of western culture. His views have inspired local winemakers:

> *Without diminishing the wonder, the winemaking process can be explained by the sciences. That appealed to me and played a large part in my decision to pursue a winemaking career. It was the appeal of making something I feel is spiritually and physically beneficial to civilization.*

Estrella River Winery, with twenty-six thousand square feet of winemaking space, was designed by Gary Eberle. This modern winery was built at a cost of $2 million in 1977 and housed the first modern laboratory with state-of-the-art equipment. *Courtesy of Gary Eberle.*

The alchemy and artistry of winemaking continue to fuel Tom's passion for making wine. He is the expert on the science of winemaking in San Luis Obispo County.

Niels Udsen

Niels Udsen and his wife, Bimmer, launched the Castoro brand in 1983 before planting organic vineyards and building the largest family-owned winery in the county. *Castoro* is the Italian translation for beaver, Niels's childhood nickname. *Courtesy of Julia Perez.*

Tom Myers mentored Niels Udsen and scores of other local winemakers on grape varieties and winemaking. Niels was hired at Estrella River Winery to work the harvest and the production line in 1981. Niels has a sharp business sense; he provided suggestions to streamline the winery's process. He developed a close friendship with Tom Myers, who introduced him to the custom crush services provided to local winemakers who purchased Estrella wine by the barrel. Niels had farmed tomatoes in Italy for a year after graduating from high school. He was intrigued with grape growing and winemaking. Tom helped Niels file the paperwork necessary to establish his own label, Castoro Cellars, in 1983.

A few years later, Niels and his wife, Bimmer, not only established the first mobile bottling service owned and operated in San Luis Obispo County, but they developed the custom crush business that is still an integral part of the Castoro Cellars business model. Niels crushed Zinfandel grapes he had sourced locally and shipped the juice to Fetzer, which was famous for White Zinfandel in the early days. Francis Ford Coppola and San Antonio Wineries became clients of Castoro Cellars. Additional clients included first-time winemakers and companies like Trader Joe's.

Niels developed his business model to offer custom crush services before he planted his vineyards and built a winery—an unusual strategy. This enabled him to acquire the capital he needed to invest and buy land for his own vineyards and winery, both of which he planned to expand gradually. Castoro Cellars is now the largest family-owned winery in San Luis Obispo County. Its vineyards were among the first to be certified as organic, a ten-year-long process. Niels later purchased the property where the Castoro Cellars Tasting Room is located, surrounded by vineyards, from George and Tommie Mulder, early growers and winemakers from the 1970s. Niels's sons, Luke and Max Udsen, founded their Bethel Road distillery in 2016. The two generations now work together.

THE SOUTH COUNTY: EDNA VALLEY VINEYARD, THE LAUNCHING PAD FOR WINE LEGENDS

Norman and Carolyn Goss: The First to Plant Chardonnay in the Edna Valley

Norman Goss holds the earliest credentials for planting Chardonnay vines in the Edna Valley viticultural region. He was first introduced to San Luis Obispo County wine by renowned concert pianist Ignace Paderewski. As a young cellist, Norm tasted his first Paso Robles Zinfandel, crafted by Paderewski himself. Goss pursued his first career as a cellist and later opened the Stuft Shirt Restaurant in Pasadena, California, in 1941, later

Norman Goss (*center*), the owner of Chamisal Vineyards, with two friends at Saucelito Canyon Vineyard in Arroyo Grande Valley. *Courtesy of Clay Thompson.*

adding additional locations in Newport Beach and Upland. He traveled extensively in Europe; the cuisine he presented was influenced by the food and wines he enjoyed abroad. By the late 1960s, Norm and his clientele were enjoying quality California wines. He auctioned off all his imported wines to concentrate solely on California vintages. The Stuft Shirt became the first restaurant in Orange County to present California wines on the menu, listed and organized by variety.

By the late 1960s, Norm noted his clientele was favoring white wine over red. He decided to research the grapes grown in California that produced these wines. Norm discovered the Edna Valley with his wife, Carolyn, in the early 1970s. They decided to purchase fifty-seven acres and plant their first vineyard. They named the vineyard Chamisal after the native, white-flowered Chamise plant that grows on the property.

Uriel Nielsen: Professional Vineyard Management

Uriel Nielsen was the first vineyardist to establish grapevines in both the Edna Valley and the Santa Maria Valley. In 1964, he planted one hundred acres of White Riesling, Cabernet and Chardonnay grapes near the Sisquoc River and Tepusquet Creek in Santa Barbara County.

Norm and Carolyn Goss hired Uriel to plant thirty thousand vines in February 1973. They selected two varieties: Cabernet Sauvignon on ten acres and Chardonnay, the Wente clone, on forty-seven acres. It was a bold experiment. The Cabernet Sauvignon vines failed to thrive because of the coastal climate; with its cool ocean breezes, it was not warm enough for the Cabernet to ripen. The Chardonnay, however, excelled. In 1976, the first vintage of Chamisal grapes was purchased and produced by David Bruce Winery and Roudon-Smith of Santa Cruz.

Jack Niven and Jim Efird Establish Paragon Vineyard Co.

Purity Grocery Store Chain in Northern California was founded by John Niven, and it was based on a unique concept that it would replace the "mom-and-pop" corner market. The spacious grocery stores featured a wide selection of fresh produce, meat and staples at 180 separate locations.

Purity Market, founded in 1929, grew to be one of the largest grocery chains in California. The Niven family rewrote wine history, as they were among the first to offer California wines at their markets, featuring Gallo and Sebastiani, among others. Wines were suddenly within reach on a daily basis, and they became part of the everyday dining experience.

By the late 1960s, a new wave of supermarket expansion created competition as larger retailers entered the grocery industry. The Niven family sold its stores one by one. Jack Niven, the son of John Niven, and his wife, Catharine, began researching new investment opportunities, focusing on the wine industry. Jack hired professors Albert Winkler from UC Davis and Vincent Petrucci from Fresno State University to identify wine-growing areas on California's Central Coast. Jack eventually selected the Edna Valley, a dairy community, on the advice from Petrucci and a handful of studies completed by UC Davis, Cal Poly and local county agricultural advisor Jack Foote.

Foote played an instrumental role in proving the success of Edna Valley as a premium wine-growing region by planting a large experimental vineyard with several grape varieties on the Righetti Ranch in 1968. By 1972, when

The first vineyards were planted in the Edna Valley in 1973. This is a photograph of the Edna Valley before hundreds of acres of vineyards were planted there in the early 1970s. *Courtesy of the Niven family.*

the Department of Viticulture and Enology at UC Davis made wine from the first harvest of these grapes, all agreed that grape cultivation in the Edna Valley was worth pursuing.

Jack Niven hired industry professionals to guide him in his new business venture. He selected 542 acres in the Edna Valley and named his business Paragon Vineyard Co. He hired Jim Efird, a recent graduate from Fresno State, to supervise the planting and development of the vineyards. Jim went to work on April 1, 1973, and began planting six weeks after vineyard manager Uriel Nielsen planted Chamisal Vineyard. Over half of the Paragon Vineyard plantings were white varieties, including Chardonnay, White Riesling, Sauvignon Blanc, Gewurztraminer and Chenin Blanc; 160 acres were planted with Cabernet Sauvignon, Pinot Noir, Zinfandel, Gamay and Merlot.

This "shotgun" approach to vineyard planting, as Jim Efird described it, expressed a leap of faith by Jack Niven and Norman Goss, pioneers of the Edna Valley. They were looking to the European wine consumers as their guides in varietal selections to market because no one had farmed these varieties in the region before. The Nivens soon concluded that Cabernet, Merlot and Zinfandel varieties could not compete with their outstanding Chardonnay, Sauvignon Blanc and Pinot Noir in the cool climate of the Edna Valley, and they made the admirable decision to graft these varieties.

Jim Efird: Professional Vineyard Management and Irrigation

Jim Efird studied agricultural economics at Fresno State and enrolled in practical courses in viticulture, soils and irrigation management. Drip irrigation was just being introduced in the plant science division at the university. Jim graduated with a new generation of viticulturists who would focus on the science and the professional management of vineyards.

Unfortunately, Paragon Vineyard's property had a limited water supply. Rather than scale back the vineyard planting by 80 percent due to the lack of water, Jim researched and installed drip irrigation on the steep hillsides, which presented new challenges. The high mineral content of the region's coastal groundwater began plugging the emitter's small passageways. Another surprising challenge was that the system had been engineered to

Pioneers of the Edna Valley (*from left to right*): Gary Mosby (winemaker at Edna Valley Vineyard), Jack Niven (owner of Paragon Vineyard) and Jim Efird (grower and vineyard manager of Paragon Vineyard). *Courtesy of Jim Efird.*

supply just enough water for one emitter per vine, which was attached to a hose in the ground. As the days grew warm and the nights grew cold, the emitter hose would expand and contract, moving the emitter several feet away from where a new vine had been planted. Jim resorted to hand watering the young vines until he was able to find a viable solution.

In the winter of 1974, Jim traveled to both South Africa and Israel, the leaders in drip irrigation technology. Jim was introduced to a new drip irrigation system by Netafim, and he replaced his existing irrigation with new emitters. The Netafim system, coupled with advanced water treatment and filtration, proved to be the silver bullet in efficiently managing irrigation on the hills of Edna Valley.

The first harvest at Paragon Vineyard occurred in 1977. Jim assumed the responsibility of selling the fruit. Since there were no production facilities located in the Edna Valley at this time, he approached small, high-quality wine producers in Napa, Santa Cruz and Sonoma Counties. The first wines from the Edna Valley grapes were made at David Bruce, Roudon-Smith, Ahern, Leeward, Glen Ellen and Felton-Empire Wineries.

The 1977 vintage was the first label to list the Edna Valley as the place of origin for the grapes. The wines made from Chardonnay grapes grown in the Paragon Vineyard were well received.

The pioneer growers of the Edna Valley, Norman Goss and Jack Niven, successfully planted hundreds of acres of vineyards in an area that was new to grapevines, but selling the grapes proved challenging. These growers combined forces to attract large producers and wineries to purchase grapes from the Edna Valley.

Dick Graff and Phil Woodward: Chalone Vineyards

Dick Graff and Phil Woodward, the founders of Chalone Vineyards in Monterey County, were introduced to Jack Niven in 1977. Chalone's 1974 Chardonnay made history at the Paris tasting in 1976. However, Chalone had experienced several years of drought, reducing its vineyard yields. It had a large production facility to fill, and it needed additional fruit to purchase. Dick Graff began buying Chardonnay and Pinot Noir grapes from Jack Niven and Paragon Vineyard Co.

Chaparral: The First Private Label for Wine Shops

In 1977, Chalone made a private label, Chaparral, for California wine retailers such as the Duke of Bourbon in Canoga Park. David and Judy Breitstein, the owners of the Duke of Bourbon in Canoga Park, were investors in the Chaparral label. Judy named the wine after being inspired by the native plants in California. The Chaparral label garnered a wide range of positive press from wine writers and great wine scores. Chalone continued the partnership with subsequent private labels, purchasing fruit from Paragon Vineyard Co. in 1978 and 1979.

Andy MacGregor: Old Vine Chardonnay Vineyard

Andy MacGregor was the third legendary grower in the Edna Valley. Andy retired from the aerospace industry in 1972; he was an engineer and part of the jet propulsion pump team that designed jet and rocket engines for the Mercury and Apollo missions to the moon. He moved to the Edna Valley with his wife, Liz, and planted twenty-five acres of Pinot Noir grapevines in 1974. Within a few years, Andy had grafted the majority of vines over to the historical Wente Chardonnay clone. Today, it is one of the oldest commercially producing Chardonnay vineyards in the Edna Valley. The old vine Chardonnay provides intensely concentrated flavors due to its small yield and deep root system, and it is now managed by Wolff Vineyards.

Margaret and Meo Zuech: Rare Italian Varieties

Romeo (Meo) Zuech met Andy MacGregor while working in the aerospace industry, and the two became lifelong friends, conducting scientific experiments together in the vineyards. Meo was an avid home winemaker who planted a small backyard vineyard and converted the family's laundry room into a small winery. He purchased wine grapes from Andy MacGregor in the Edna Valley and Benito Dusi in Paso Robles. Margaret and Meo eventually sold their home and moved onto Andy's property in the Edna Valley to help manage the vineyards.

Meo was interested in two grape varieties that were grown in his native Italy. The Lagrein and Teroldego varieties were imported to America by Meo, in partnership with UC Davis, from the Alto Adige region of Italy. The varieties were grafted on vines in the Edna Valley with Andy MacGregor's assistance. Today, the Edna Valley is home to the largest Lagrein vineyard in California.

Above: (*From left to right*) Liz and Andy MacGregor with Jack Foote, circa 1980. Jack Foote, the San Luis Obispo County agricultural advisor, planted experimental vineyards throughout the county in the 1960s and 1970s. His research, funded by the University of California, encouraged growers like Andy and Liz MacGregor to plant vineyards in 1975. *Courtesy of Clay Thompson.*

Left: Meo Zuech started making wine in the laundry room of his home in Westlake Village in the 1970s. Meo and his wife, Margaret, were featured in a 1978 *Sunset Magazine* article that described their home winemaking techniques. *Courtesy of the Zuech family.*

Chuck Ortman: Mr. Chardonnay

Chuck Ortman left a career as a graphic artist to become a cellar rat in Napa Valley for winemaker Joe Heitz in 1968. He learned the craft of winemaking just as the California wine revolution freed winemakers to create their own unique styles. Chuck made wine for several revered wineries in the Napa Valley, including Spring Mountain Vineyard, Far Niente, Shafer and St. Clement. He focused on Chardonnay and became known as the master of Chardonnay, in part because of his experiments and innovations with barrel-fermented Chardonnay. Chuck was the first winemaker to develop this fermenting process in the early 1970s, and it quickly spread across California, becoming the Chardonnay of choice for the American consumer.

He founded his own enology firm and purchased grapes from the Edna Valley to make his highly praised 1979 Chardonnay. Chuck founded Meridian Wine Cellars, in 1984, its name a nod to Chuck's love of sailing. *Meridian* means "the point or period of highest development." In 1988, Napa Valley's Beringer Vineyards acquired Meridian with Chuck leading the helm of winemaking operations out of the former Estrella River Winery in Paso Robles. After fifteen years and growing Meridian's production to over one million cases a year, Chuck retired and began a new family venture in the Edna Valley with his son Matt: Ortman Family Wines.

Bill Greenough Rediscovers Henry Ditmas's Saucelito Canyon Vineyards

In 1974, William Greenough rediscovered the Saucelito Canyon Vineyards of Rosa and Henry Ditmas in the Upper Arroyo Grande Valley and spent the rest of the decade bringing the original Zinfandel vines back to life with his pick and shovel. As the vineyard of ancient gnarled vines revived, Bill head-pruned the dry-farmed vines and looked forward to his first harvest. Bill had started winemaking with grapes he had harvested in the Zinfandel vineyards of Mel Casteel, where he and his friends had sometimes stomped grapes in the nude among the vines. (It was the 1960s winemaking style.) Bonded in 1982, Bill began making his Zinfandel and Cabernet wines. Bill described his vineyard as having a lot of strength and character, a vineyard whose roots reach deep into Arroyo Grande Valley soil and history. The winery is run by Bill's wife, Nancy; his daughter Margaret; and his son Tom, who is now the winemaker.

Above: Saucelito Canyon Vineyard, circa 1974. Bill Greenough acquired the property and discovered the abandoned Zinfandel vineyards, which had been planted by Henry Ditmas in 1880. *Courtesy of the Greenough family*.

Left: Second-generation winemaker Tom Greenough handcrafts Zinfandel for Saucelito Canyon after being mentored by his father, Bill Greenough, circa 2018. *Courtesy of the Greenough family*.

Historic Sites to Visit

Edna Valley Vineyard
Saucelito Canyon Winery
The town of Old Edna
Biddle Ranch and Corbett Canyon Road
Chamisal Vineyards
Center of Effort Winery
Baileyana Tasting Room and the Historic Schoolhouse
Wolff Vineyards

8

The 1980s

The AVAs and New Legends in Winemaking

GRAPE VARIETIES: CABERNET SAUVIGNON

Introduction

Commercial wine growing and production increased dramatically in the 1980s in San Luis Obispo County. Large producers and winemakers from established viticultural communities in Northern California shared their knowledge and technology. This created new opportunities and innovation for local growers; many new wineries were bonded and built new production facilities throughout the county. The pioneering spirit of the growers in Paso Robles and the coastal region of San Luis Obispo County to develop a regional identity led to the founding of three American viticultural areas (AVAs) in the 1980s: Paso Robles, York Mountain and Edna Valley.

North County: Paso Robles AVA and York Mountain AVA

The passion of new wine lovers in the 1960s and 1970s was focused on the classics: Cabernet Sauvignon and Chardonnay. Cabernet is known for

Paso Robles wine country with the iconic oak tree growing in the vineyard. *Courtesy of Paso Robles Wine.*

its complex bouquet and flavor. Chardonnay is the grape that produces some of the finest dry white wines in the world. In San Luis Obispo County, Cabernet Sauvignon gradually replaced Zinfandel as the leading wine produced in the Paso Robles area, while Chardonnay from the San Luis Obispo Coast region continues to receive accolades. Both the northern and southern regions of the county enjoy Mediterranean climates, producing premium grapes.

Major investments by large wine companies such as J. Lohr have helped fuel the growth in San Luis Obispo's vineyard acreage and brought world-class winemakers to the county. The Central Coast AVA, established in 1985, covers multiple regions and terroirs in six counties ranging from San Francisco to Santa Barbara. The defining characteristic of the Central Coast is its cooling influence from the Pacific Ocean. Close to 50 percent of the vineyard acreage in the county is planted with Chardonnay.

Cabernet Sauvignon is the leading variety in Bordeaux; the Haut-Médoc region is the most famous growing region. Professor Carol Meredith at UC Davis has found strong DNA evidence to suggest that Cabernet Franc and Sauvignon Blanc may be the parents of this popular grape. It was first imported to California in 1852, but most vineyards died out during the Prohibition era. By 1940, there were only about 400 acres of the grape left, but this number rose to 600 acres in 1960. When the California wine revolution took hold in the 1970s, this acreage rapidly rose to 26,742. Cabernet Sauvignon became the red varietal that wineries needed to produce in order to establish their credibility and brand. The Judgment of Paris in 1976 was the blind tasting at which the 1973 California Cabernet made by Stag's Leap Wine Cellars in Napa won best over French producers. This stunning result brought worldwide attention to California. Wine lovers and producers began seeking out California wines.

The Cabernets produced in the Paso Robles area were impressive. The biggest obstacle these winemakers had was the marketing and selling of local wines to new markets in California and throughout the United States. Most outsiders thought wine made in Paso Robles was produced somewhere in the state of Texas. A small group of men and women began to devise a long-term marketing strategy that would include a wine festival that became one of the largest in the United States.

Winery owners Tom Martin and Gary Eberle; Victor Hugo Roberts, a winemaker; and grower Herman Schwartz worked together to establish Paso Robles as a well-known wine region and tourist destination. The decision to establish the Paso Robles AVA was crucial to their success. Tom Martin coordinated the application process in his office. Each of the others recruited the support of growers and winery owners and summarized the research needed to complete the necessary documents. When it was established in 1983, the Paso Robles AVA was known as the "largest un-subdivided" AVA—with approximately 614,000 acres—famous for its heritage varietal Zinfandel. At that time, there were seventeen wineries in the Paso Robles area and 5,159 acres of grapes in production in all of San Luis Obispo County. At the time, vineyard land was selling in the range of $3,000 to $7,000 per acre, much more reasonable than the $30,000 to $40,000 range in the famous Napa Valley.

It was said that the cost of bringing a vineyard into production was $10,000 per acre. This meant that over $50 million had been invested in vineyards in 1983. This local investment was generating over $10 million of annual revenue in the county. Suddenly, local governments realized the importance of promoting the grape and wine industry. A major economic shift occurred. Many new growers bought land and planted vineyards. Over two hundred wineries were established over the next two decades. Plans were made to build new hotels and open gourmet restaurants, forming the foundation of a growing hospitality industry.

The Paso Robles AVA is notable for its geographical diversity; it includes the Estrella Plains, the Templeton Gap and the rolling Adelaide Hills. The diversity of the terroir continues to provide wine lovers with over fifty varieties of vinifera grapes and a wide range of winemaking styles. By 2014, the Paso Robles AVA was subdivided into eleven sub-AVAs, each characterized by microclimate, soil, topography and annual rainfall. Both traditional dry-farming techniques and irrigation can be found in these vineyards.

In 1984, the smallest AVA in the United States was established, the famous York Mountain AVA. It is located west of Paso Robles. Its application was submitted by Max and Stephen Goldman, the owners of the historic York Mountain Winery. The vineyards of this region were planted at an elevation of 1,500 feet on the eastern side of the Santa Lucia Mountains. The area takes up just 14.62 square miles, comprising 9,360 acres.

Welcome to Paso Robles: The First Annual Wine Festival in the Park

The Paso Robles Wine Festival was founded in 1983 to raise funds for the Paso Robles AVA application and to bring tourists to the city park in the heart of downtown Paso Robles. Seventeen wineries set up booths and poured their wines. Eleven winemakers served varieties of Heritage Red, Rosé and White Zinfandel. The sixteen original wineries were Martin Brothers, Caparone, Ranchita Oaks, HMR Ltd., Farview Farm, Creston Manor, Tobias, Estrella, Eberle, Mastantuono, Pesenti, Old Casteel, Las Tablas, York Mountain, Watson and Twin Hills. Only two are still in production: Eberle and Caparone wineries. The others either have new owners, new names or have had totally new wineries established in their historic vineyards.

This was one of the first annual wine festivals in California, and it grew to be the largest in the nation, with thousands of attendees, including food and artisan purveyors. The visitors to this wine festival were originally from Southern California, but now, they come from all over the world. The wine festival has expanded to cover a large number of events at wineries and local venues. It includes seminars, tastings, concerts, films, exhibits and wine dinners.

New Wine Legends Emerge in Paso Robles

Justin Baldwin

Justin Baldwin grew up in San Francisco, studied business and entered the banking industry. He worked in large cities, including London, New York and Los Angeles. He acquired a passion for good wines and food, and when he entertained wealthy clients, he selected fine wines to share with them. His favorites were the wines of Bordeaux. He watched others pursue their dreams of planting vineyards and making wines, and he decided to start pursuing his own passion for winemaking.

As a businessman, he made the financial decisions that would lead to developing a profitable vineyard and winery. He started looking for land where he could plant his own vineyard. He worked in Los Angeles, so he purchased land he could reach within four hours. This way, he would be able to work in the vineyards on weekends. He purchased 160 acres on Chimney Rock Road in Paso Robles from actor Gary Conway in 1981.

Justin Baldwin, the founder of Justin Vineyard and Winery, is known for his Isosceles label. *Wine Spectator* named this wine one of the top ten in the world in 2000. *Courtesy of Julia Perez.*

Justin and his wife, Deborah, built a winery and hired winemaker Steve Glossner. Their first harvest in 1987 produced the Justin Reserve Wine, which won the Sweepstakes Award at the Los Angeles County Fair in 1988. The wine was relabeled Isosceles. In 1997, Justin's 1994 Isosceles won the Pichon Longueville Comtesse de Lalande trophy for the "Best Blended Red Wine Worldwide." The competition was held by the International Wine and Spirit Organization in London, which had hosted the competition that had been won by Stanley Hoffman and his 1973 HMR Chardonnay.

Justin sold his winery in 2010, and the Baldwins went their separate ways. Justin continues to travel and hosts over 150 wine events per year. He has been a tireless advocate for Paso Robles wine.

Doug Beckett

Doug and Nancy Beckett moved to Paso Robles from San Diego County to raise their sons in a healthy rural environment. Doug bought a ranch with a walnut grove on Peachy Canyon Road in 1982; with a master's degree in educational psychology and a strong business background as the owner of a chain of convenience stores, he wanted to be a winemaker. He met the legendary grower Benito Dusi and winemakers Tom Myers and Ken Volk; all three served as his mentors for growing and making the wine of Zinfandel grapes. As an artist, he considered winemaking an art form.

In 1983, Doug was introduced to Pat Wheeler, who became his partner in Tobias Winery. Pat introduced Doug to local winemakers, including the Pesenti, Nerelli and Rotta families; the Goldmans from York Mountain; and Niels Udsen, the founder of Castoro Cellars.

Doug was a founding member of the first Paso Robles Wine Festival, where he poured Tobias 1981 Zinfandel and 1982 White Zinfandel. Doug planted his first vineyard using the budwood from the famed Benito Dusi Vineyard. He opened his first tasting room in Cayucos and the second at his own Peachy Canyon Winery in 1989. Doug has been a tireless promoter of the Paso Robles area ever since.

Doug and Nancy Beckett, the founders of Peachy Canyon Winery, raising a glass of Zinfandel with Tom Westberg in 1995. *Courtesy of Dan Hardesty.*

Doug became an award-winning Zinfandel winemaker early in his career. He was on the 1991 and 1992 *Wine Spectator* list, recognized with the top two Zinfandels for two consecutive years. Peachy Canyon was the first winery in the Paso Robles area to receive this recognition. He has sold Peachy Canyon wine in Europe, Asia, Mexico, Canada and every state in the United States.

Doug became a founding member of Zinfandel Advocates and Producers (ZAP) in 1991. When he became the president of ZAP in 1999, he supported the "saving of old vine Zinfandel and heritage vineyards," a project that has since become highly regarded in California. He was the cofounder of the Zinfandel Festival in Paso Robles, and he presented Paso Robles wines in Cuba as a member of the California trade group in 2015, when the U.S. and Cuba trade embargo was lifted.

The Beckett family has owned Peachy Canyon Winery for over thirty years. Doug and Nancy's sons, Jake and Josh, who founded and sold the famous Chronic Cellars, have now joined them as the second generation of Becketts managing the vineyards and directing the winemaking at Peachy Canyon Winery.

Max Goldman

Max Goldman started his career as a chemist at a Lodi winery in 1934, but he quickly embraced winemaking. His career spanned four decades of the most exciting growth and development in the California and New York wine industry. He observed and recorded the changes in the industry and collaborated with professors at UC Davis and Fresno State to develop the techniques and standards of quality winemaking. He dramatically improved the fermentation techniques in

Max Goldman purchased York Mountain Winery in 1970. He replanted the vineyard, rebuilt the winery and made award-winning wines for the next thirty years, with his son Steve as the winemaker and his daughter Suzanne as the head of marketing and community relations. *Courtesy of Suzanne Goldman Redberg.*

producing wine and Champagne. He was a founding member of the Wine Institute in San Francisco in 1934, which helped establish quality winemaking in California after Prohibition. He purchased the iconic York Mountain Winery in 1970 and replanted the vineyard. There, he restored the winery and produced award-winning wines with his son Steve as the winemaker. His daughter Suzanne designed the tasting room, developed the marketing plan and partnered with local charities to produce major wine events at the winery. Max shared his expertise with local winemakers and promoted San Luis Obispo County as the next legendary winemaking area in California until his death in 2004. York Mountain Winery was sold in 2001.

Victor Hugo Roberts

Victor Hugo Roberts represents the essence of the small, high-quality producer that shaped the wine history of San Luis Obispo County in the 1980s and 1990s. He was the fourth winemaker trained in enology at UC Davis to move to the county. Victor was one of the early UC Davis graduates to settle in the area, following Gary Eberle, Tom Myers and Steve Dooley. He was an early pioneer in establishing winemaking as an art in conjunction with science.

Victor Hugo Roberts represents the essence of the small high-quality producer that shaped the county between 1980 and 1990 and continues to do so. *Courtesy of Julia Perez.*

Victor Hugo graduated in 1979, working first for the Italian Swiss Colony in Sonoma County and for Brookside Vineyards in Temecula and Rancho Cucamonga. He moved to Paso Robles with his wife, Leslie, in 1982 to become the winemaker and general manager at Creston Vineyards. His role as general manager expanded to the management of a multilevel general and limited partnership that was formed for tax shelter advantages rather than for growing quality grapes and making wine. The tax shelter legislation represented a new era, which brought limited partnerships and corporate structures to Paso Robles and the Edna Valley in the 1970s and 1980s.

By 1985, Victor Hugo knew he wanted to establish his own vineyard and winery. He now farms seventy-eight acres of his one-hundred-acre ranch located on El Pomar Drive in Templeton. He released his first vintage of wines in 1999. He once provided consulting services to local wineries but now focuses on his own brand. The Victor Hugo Tasting Room opened in a historic old barn on the property, where visitors and wine club members can gather.

Victor Hugo joined Tom Martin, Herman Schwartz and Gary Eberle in developing a marketing plan to make Paso Robles a destination and the wines of the Paso Robles AVA famous. He also served on the committee that was formed by the Paso Robles Chamber of Commerce to establish an annual wine festival in 1983. He served as the president of the Paso Robles Wine Festival for eight years as it expanded to become the largest wine festival in the United States.

Ken Volk

Ken is a third-generation California native with a talent for winemaking that is even more spirited than the name of his first winery, Wild Horse, which was established in 1982. The winery was named in honor of the descendants of the first Spanish horses brought to California. Ken expanded the range of varieties that could be produced by one man; he experimented with over thirty-one extraordinary and complex grapes sourced from vineyards located throughout California. Ken first made wine in plastic bins in a Cayucos garage after graduating from Cal Poly with a bachelor's degree in fruit science in 1981. He was hired to work the fall harvest at Edna Valley Vineyard and then discussed buying his own vineyard with his family.

Ken Volk and Archie McLaren at the Central Coast Wine Classic. Winemaker Ken Volk is famous for experimenting with thirty-one grape varieties, crafting fine wine from each. He founded his first winery, Wild Horse, in 1982. *Courtesy of Dan Hardesty.*

With his family's backing, Ken chose land in Templeton, where he planted his first vineyard in 1982. He built the winery and in 1986 released his first wines, which included 125 cases of Pinot Noir and 450 cases of Cabernet. He married chef Tricia Tartaglione in the same year. They have supported many wine dinners and events to raise money for charity and support the local wineries and restaurants in San Luis Obispo County.

Ken sold Wild Horse to Peak Wines International in 2003. He has received multiple awards, including those for agriculturist, winemaker and winery of the year in the 1990s. He continues to make wine at Kenneth Volk Vineyards in Santa Maria.

Winemaking Facilities Open in Edna Valley and Arroyo Grande Valley

The capacity to crush and produce wines in the Edna Valley grew exponentially with the addition of three new winemaking facilities. During the 1970s, the harvested fruit was sold to wineries in Napa and the Central Valley because there were no production facilities.

The Edna Valley harvest crew, circa 1982 *(from left to right)*: Gary Mosby, Tim Lloyd, Clay Thompson, Bruno D'Alfonso and Frank Focha. *Courtesy of Clay Thompson.*

Three growers decided to build their own wineries between 1979 and 1981. Lawrence Winery was constructed in 1979 with the capacity to produce up to 260,000 cases of wine. Norman Goss, the owner of Chamisal Vineyards, established a small winery in 1980 with the capacity to produce 5,000 cases. In 1981, Edna Valley Vineyard, a partnership between the Niven family's Paragon Vineyard Co. and Chalone Vineyards founded by Dick Graff and Phil Woodward, built a 25,000-case production facility, replicating the Chalone Winery, which borders the Pinnacles National Park in Monterey County, California. This new winery, Edna Valley Vineyard, was equipped with state-of-the-art winemaking technology, a scientific laboratory, a bottling room and an underground concrete cave for barrel aging.

Edna Valley Vineyard became the launching pad for new winemakers who founded their own wineries, including Ken Volk of Wild Horse, Steve Dooley of Stephen Ross Wine Cellars and Clay Thompson of Claiborne and Churchill Winery.

Chardonnay Becomes the Star of Edna Valley

By the early 1980s, Paragon Vineyard Co. was producing high-quality Chardonnay grapes that caught the attention of many California winemakers, including Greg Bissonette of Chateau Chevalier in Napa Valley, Chuck Ortman (Meridian and Ortman family) and Richard Sanford (Sanford Winery and Alma Rosa) in Santa Ynez.

Edna Valley Chardonnay's notoriety grew both nationally and internationally through three comparative blind tasting events of French and California wines in 1982. The events were hosted in San Francisco, Houston and Dallas and were successors of the famed blind wine tasting event that was held in 1976, the Judgment of Paris. The California-versus-France rivalry pitted California wines against French wines. The result: California Cabernet and Chardonnay wines from Stag's Leap and Chateau Montelena were rated the top wines, beating their French rivals in blind tastings by French judges. Out of the eight Chardonnays tasted in the blind flight, a tie for first was announced between two California Chardonnays: the 1980 Edna Valley Vineyard Chardonnay from Paragon Vineyard and the 1979 Trefethen from Napa Valley. Both Steven Spurrier and the late wine writer Robert Finigan rated the Edna Valley Vineyard the top Chardonnay wine. The Edna Valley Vineyard Chardonnay was described by Finigan in a May 20, 1982 article in the *Houston Chronicle*:

> [It is] *golden, straw colored and dry, but rich flavored with lots of wood aging coming through on the nose. It is clean and crisp with buttery overtones in the style of many popular California Chardonnays. Considering its price—about $12 a bottle—it's also number one and an excellent value.*

On the thirtieth anniversary of the Judgment of Paris, another local Chardonnay ranked number one. Simultaneous blind tasting events were held in London and Napa Valley. Arroyo Grande producer Talley Vineyards won the top honor among all California Chardonnays for its 2002 Rosemary's Vineyard Chardonnay.

A NEW AVA IS BORN

Jack Niven put forth a petition to establish the Edna Valley as an American viticultural area on September 11, 1980. Joining Niven to help prepare documents and collect information for the AVA petition were wine growers Norman Goss of Chamisal Vineyards, Jim Lawrence of Lawrence Winery and Andy MacGregor of MacGregor Vineyards. Jack submitted topography maps, soil data and San Luis Obispo County planning studies to establish the boundaries for the eleven-thousand-acre AVA. The Edna Valley was the tenth AVA to be established in 1982. This designation allowed winemakers to use Edna Valley as the region of origin of fruit and wine production on their wine labels as well as the term "estate." Jack's historical research confirmed that the name "Edna" was derived from the local Old Edna townsite that was established back in the 1800s. The townsite property has a history that can be traced to the 1840s, when it was part of a Mexican land grant awarded to Jose Maria Villavicencia.

Andre Tchelistcheff wrote to Jack Niven in support of the Edna Valley AVA, stating:

> *In my long-lasting career as a consulting enologist...I have had the opportunity to observe and study the performance of the Edna Valley*

Coastal influence defines the terroir of San Luis Obispo County Coast wine. In this photograph of the Edna Valley, fog can be seen funneling in from nearby Morro Bay, and it is protected by the rolling vistas of the Santa Lucia Mountains. *Courtesy of the San Louis Obispo Coast Wine Collective.*

> *vineyards during the last several years, from 1975 to 1980. The ecological complex of the valley soil, micro-climate and the productivity is directly and strongly reflected in the individual qualitative structure of the wines produced exclusively from the grapes grown within the boundaries of this specific viticultural area.*

The Edna Valley AVA stretches along the Santa Lucia Mountain Range to the northeast and a low hilly complex paralleling Tiffany Road to the southeast. On average, the vineyards are located five miles from the Pacific Ocean, which produces cool maritime ocean fog that blankets the valley throughout the year. The exceptionally long growing season begins with bud break in February, and harvest lasts through November. It is the coldest AVA in California.

Catharine Niven's Tiffany Hill Begins Legacy of Baileyana Wines

Jack Niven's wife, Catharine Niven, started a new venture in 1981 to plant her own vineyard and make a very personal, Burgundian-style wine. She planted a small three-acre vineyard in the front yard of the family home on Tiffany Hill Road in the Edna Valley. Her first wine, a Chardonnay released in 1984, was named Tiffany Hill. Catharine selected the name Tiffany Hill as her label, memorializing the street the family lived on in Edna Valley.

Catharine was the first woman in the Edna Valley to plant a vineyard, produce her own wine label and develop a marketing strategy to sell her wines. She sold directly to restaurants in Honolulu, Los Angeles and San Francisco by making phone calls to describe her Chardonnay. When a bottle of the Tiffany Hill Chardonnay was presented to the chairman of the famous jewelry store Tiffany & Co., he filed a "cease and desist" order. A name change was required, and the Chardonnay was renamed Baileyana, a nod to the street Jack grew up on in Hillsborough, California.

By 1989, the family began producing their wines under the new label, Baileyana. Although Catharine continued to market the wines, the Nivens needed professionals to manage Paragon Vineyard Co. Jack Niven and Jim Efird founded Pacific Vineyard Company in 1988 to manage the vineyards. George Donati joined Pacific Vineyard in 1995 and became the head of the company when Efird retired in 2004.

Old World Invests in the New World, Bringing Sparkling Wine to San Luis Obispo County

In the 1970s and 1980s, many grandes marques of Champagne established vineyards and sparkling facilities in California, including Domaine Chandon, Mumm Napa, Domaine Carneros and Roederer Estates. Champagne producer Deutz (formerly known as Deutz & Geldermann) discovered a cool-climate site south of the Arroyo Grande Valley, just off of Highway 101. It became the center of the "sparkling wine revolution" in San Luis Obispo County in the 1980s.

Andre Lallier-Deutz, a fifth-generation proprietor and chef de caves of Champagne Deutz of Ay, France, began the international search for

(*Left to right*) Eric Hickey, Lino Bozzano and Dave Hickey. Dave Hickey started his career as an electrician at Maison Deutz and spent over twenty years as winemaker for Laetitia Winery. Eric Hickey is now the vice president for Laetitia. *Courtesy of Laetitia Vineyards and Winery.*

locations to produce sparkling wines in the tradition of the House of Deutz in 1978. Three years later, Andre visited the Arroyo Grande Valley; it was located three miles from the Pacific Ocean and had rolling hills, gravelly low-yielding soils and cool climate conditions very similar to the area where Champagne Deutz was founded in France. Andre purchased the property and named the new winery Maison Deutz. In 1982, the 185-acre vineyard in Arroyo Grande was planted equally with Pinot Noir, Chardonnay and Pinot Blanc grapes, the three main grape varieties for classic Champagne.

Harold Osborn was hired as a winemaker after working at Schramsberg Vineyard in the Napa Valley for eight years. The new production facility mirrored the original Champagne Deutz gravity flow design, and the building was completed in 1983 with the intention of creating sparkling wines by using the traditional méthode champenoise. This method calls for hand harvesting and the use of a traditional basket press among other protocols when producing sparkling wine.

Maison Deutz became the leader in méthode champenoise, and it was the only winery in California and the western hemisphere to operate two French Coquard basket presses. Dave Hickey, an electrician and contractor for the winery, was hired on to help with the facility's maintenance. When the 4.4-ton Champagne basket presses arrived from France, Dave Hickey and Andre Lallier-Deutz worked together to assemble them in preparation for harvest.

Christian Roguenant, the assistant winemaker at Champagne Deutz in France for two years, was asked by Andre Lallier-Deutz in 1986 to become a winemaker at Maison Deutz after Osborn's departure. Roguenant jumped at the opportunity to produce "New-World" wines in Arroyo Grande, California. Christian hired Dave Hickey to continue assisting in sparkling wine production. Dave was mentored by Roguenant for the next twelve years.

Christian left Maison Deutz and spent the next twenty years working for the Niven family and crafting an array of Burgundian and Aromatic varietals. Dave Hickey became the next sparkling winemaker at Maison Deutz. The winery was sold and renamed Laetitia in 1994. A few years later, in 1997, the winery was sold again. Dave managed the sparkling wine program for Laetitia for twenty years before retiring in 2019. Eric Hickey, his son, is now the president of Laetitia Vineyard and Winery.

From Farming to Vineyards: The Arroyo Grande Valley

What started as a family farm in 1948 has now grown to become a major agricultural supplier of both vegetables and fine wines. Oliver Talley established his vegetable farm in the Arroyo Grande Valley, growing broccoli, beans, cauliflower, peppers and tomatoes. In 1963, Oliver's son Don returned to the farming business after graduating with a degree in agricultural business management from UC Berkeley in 1962. He married Rosemary Green that same year and began growing the family farm with new business initiatives.

Don incorporated Talley Farms in 1966, and the company began purchasing the farmland that his father had historically leased. Don, the president of Talley Farms, developed new management practices and technological advances, including the construction of vegetable coolers. By the early 1970s, the Talley family had purchased 150 acres of farmland in the Arroyo Grande Valley, along with the El Rincón Adobe and ranch. The historic El Rincón Adobe was built in the early 1860s by Ramon Branch, the son of one of the original settlers of the Arroyo Grande Valley. The adobe

Oliver Talley, the founder of Talley Farms. *Courtesy of the Talley family.*

Don Talley and Vicki Carroll raising a toast at the harvest celebration in the Edna Valley. *Courtesy of San Louis Obispo Coast Wine.*

was built with sun-dried mud bricks crafted on site from local soil. It has been restored and now houses the offices of Talley Vineyards.

The land surrounding the Rincon adobe is located on steep hillsides. Don realized it would be impossible to farm traditional crops there, so he researched other alternatives. Farmers were planting vineyards in the Santa Maria and Edna Valleys. Don and Rosemary Talley were the first to plant grapes in the Lower Arroyo Grande Valley. Don consulted with viticulturists from UC Davis and planted Chardonnay and Pinot Noir grapes in the Rincon Vineyard next to the adobe in 1982. He selected the Wadenswil 2A clone for an optimal Pinot Noir. This disease-resistant clone, originally grown in Switzerland, excels in cool climates. Don and his wife, Rosemary, originally planned to sell their grapes to local winemakers. At the first harvest in 1986, they were impressed by the quality of the grapes and decided to start their own winery. They hired winemaker Steve Rasmussen, who had worked at the Mondavi properties in Lodi and Oakville, and soon began making award-winning Chardonnay and Pinot

Noir. Rosemary, who was head of produce sales at Talley Farms, managed the winery business.

The Rincon Adobe opened as a tasting room in 1987, featuring wines from the Talley and Greenough families. The two wineries shared a unique partnership in the tasting room. Talley Vineyards offered tastings of white wines and Saucelito Canyon presented tastings of Zinfandel in the historic adobe. Both wineries soon built their own tasting rooms in separate locations. The new Talley tasting room overlooks the vineyards, and the Saucelito Canyon tasting room is located on Biddle Road in the Edna Valley. Don's son Brian Talley now heads the winery's operations. The Talley Estate Chardonnay was recognized by *Wine Spectator* as one of the top one hundred wines in the world in the year 2000, and the 2012 Rosemary's Vineyard Pinot Noir was awarded a score of 98 in Robert Parker's *Wine Advocate.*

A Second AVA in South County

Growers Don Talley and Bill Greenough saw the potential in creating a new American viticultural area; their vineyards were not included in the Edna Valley AVA. Over Don's coffee table, the two developed an AVA petition, researching soils, climate and boundary conditions. The Arroyo Grande Valley AVA was established in 1990, along with a new marketing association that was dedicated to promoting both coastal regions of San Luis Obispo,

Meo Zuech, Nancy and Bill Greenough, Margaret Zuech and Steve and Paula Dooley were some of the first board members of the Edna Valley Arroyo Grande Valley Vintners and Growers Association. *Courtesy of the San Luis Obispo Coast Wine Collective.*

the Edna Valley Arroyo–Grande Valley Vintners and Growers Association. Steve Dooley, the winemaker at Edna Valley Vineyards at the time, was the first president of the association's board, along with second-generation vintner John R. Niven of Baileyana Winery.

Historic Sites to Visit

Paso Robles Wine Festival
JUSTIN Vineyards and Winery
Peachy Canyon Winery
Victor Hugo Winery
Kenneth Volk Vineyards
Vintage Paso Zinfandel Weekend
Talley Vineyards
Stephen Ross Wine Cellars
Laetitia Vineyard & Winery
Oceano Dunes Natural Preserve

9

The 1990s

Rhone Varieties and the Rhone Rangers

GRAPE VARIETIES: GRENACHE, SYRAH, MOURVEDRE, CINSAUT, ROUSSANNE, VIOGNIER, GRENACHE BLANC, COUNOISE, PICPOUL BLANC, CINSAUT, BOURBOULENC, VACCARÈSE

Introduction

Kermit Lynch, a San Luis Obispo native, opened his wine shop, Kermit Lynch Wine Merchant, in Albany near UC Berkeley with a loan from a friend. He imported wines from France, traveled to France with cookbook author Richard Olney and began importing wines from the Rhone Valley in southern France. The Rhone River is over five hundred miles long and spans the area between the Rhone Alps and the ancient city of Arles. Grenache grapes are grown in this region.

In the 1880s, Rhone grape varieties were grown in San Luis Obispo County. This included the red, Grenache, Mataro, Cinsaut and Carignane, and white varieties, Marsanne and Roussanne. In the twentieth century, Rhone varieties were grown in the Central Valley and were used by producers for blending. But the traditional Rhone grapes—Grenache, Mourvedre and Viognier—that had been planted in California in earlier decades were dying out by the early 1980s. Growers replaced them with new varieties to capitalize on the demand for the classics, Cabernet Sauvignon and Chardonnay, and popular white wines such as Sauvignon Blanc and Riesling.

Gary Eberle, the winemaker at Estrella River Winery, had planted the first Syrah in the 1970s and produced an award-winning 100-percent Syrah varietal, although he had trouble marketing it. He abandoned his initial efforts to produce the wine, but he did make the Syrah clone, which he had obtained from a UC Davis test vineyard, available to Bob Lindquist and other growers who were interested in reviving the Rhone varieties. Gary Eberle and Ken Volk, the owner of Wild Horse Winery at the time, were early Syrah enthusiasts.

In the 1980s, a group of winemakers, including Randall Grahm, Joseph Phelps, John MacCready, Bill Crawford, Fred Cline, Robert Lindquist and Steve Edmunds, began experimenting with Rhone varieties instead of focusing on Chardonnay and Cabernet. During that same decade, John Alban was the pioneer in saving Viognier from extinction in California and in researching Rhone varietals in France. Gary Eberle sold many of the cuttings to Bob Lindquist, who became famous for biodynamic farming and his Syrah under the Qupé label.

Bob Lindquist in his biodynamically farmed vineyards. *Courtesy of Fran Collin.*

Robert Haas, an American wine importer along with the Perrin family, the owners of Chateau de Beaucastel, formed a partnership to establish a vineyard growing Rhone varieties in California. They spent several years searching for the location that met their criteria for terroir. The Adelaida District, just west of Paso Robles, was their choice in 1989; the partners named their vineyard Tablas Creek. This area of Paso Robles provides the ideal terroir for these varieties.

The vineyards and wineries of Austin Hope, Justin, L'Aventure, Linne Calodo, Saxum and Summerwood followed the lead of Tablas Creek and winemaker Neil Collins in producing Rhone varietals in the twenty-first century. Since 1997, there has been a dramatic increase in the California acreage that is planted with Rhone varieties, and many of these Rhone vineyards are located in San Luis Obispo County because of its multiple microclimates. The Rhone movement inspired supporters to create two nonprofits. Both organizations have supported growers internationally and locally.

Mat Garretson organized the Viognier Guild, the predecessor to Hospice du Rhone. By 1999, John Alban had changed the name to Hospice du Rhone, which he felt allowed all the growers and makers of Rhone varieties to feel welcome and part of the organization. Vicki Carroll joined the team and developed the organization as an international vintners' association and resource for Rhone growers and producers worldwide. This organization is headquartered in San Luis Obispo County.

A second organization, the Rhone Rangers, established local chapters to promote the Rhone varieties produced by their members. It provides educational materials to the public and their members. The San Luis Obispo County chapter is the largest in the nation with over fifty members.

The John Alban Story

John is one of the most influential American Rhone producers. John noted that the premier wine areas will endure and continue producing wine five hundred years from now. He spent four years in the 1980s studying Rhone varieties in areas of Condrieu, France, which is known for its Rhone vineyards. He returned with vine cuttings.

John realized that none of the other areas producing Rhone wines in the world had the critical mass or resources that were needed to successfully market their Rhone wines in the 1980s. He began sharing his knowledge as a consultant in 1985. His nickname "Johnny Appleseed" stuck. As a consultant, John had a nursery business selling the cuttings of Rhone varietals that he selected for California winemakers.

John Alban, one of the most influential growers and producers of American Rhone and the pioneer of the Rhone movement. *Courtesy of Hospice du Rhone.*

In 1989, John found the perfect place, sixty acres on the three-hundred-acre property in the Edna Valley, to plant Syrah, Viognier and Roussanne. First, he planted a thirty-two-acre block of Viognier. By doing so, he said, "I almost single-handedly doubled the world's acreage." Alban is a pioneer in promoting Viognier, as the decline in the acreage planted to this Rhone variety almost doomed it to extinction.

John founded the first American winery and vineyard established exclusively for Rhone varieties. His label on the back of every Alban wine bottle emphasizes that point. The land where Alban planted these vines had no prior wine history. Its soil included the hoofprints of the cattle and dairy industries. The steep hillsides are at a higher elevation, covered with chalk and limestone soils. On this terroir, John grows vines with small clusters of berries cooled by ocean breezes and morning fog. The Edna Valley's growing season is long, which results in greater texture, color and flavor in the grapes. The grapes are harvested at night to produce more fruit flavor.

When John poured his 2013 Alban Estate "Reva" Syrah from the Edna Valley, he described the wine in the following terms: "This wine is never shy or dainty, but you may be more surprised by its grace than its girth." *Wine Spectator* gave the wine a rating of ninety-four points. The wine was aged for three and a half years for polish and layering. Other wines he produced included the Lorraine Reva Vineyard Syrah, Seymour's Vineyard Syrah and Pandora, a blend of Grenache and Syrah. He also produces a dessert wine with 5 percent Botrytis grapes aged in 50 percent new oak.

Hospice du Rhone

John met Mat Garretson in Piedmont, Georgia, at his initial Viognier Guild meeting in 1991. There was an audience of twenty people and thirty-five Viognier and Condrieu wines to taste. John was one of two producers who attended this event, which Mat characterized as the largest tasting of domestic and imported Viogniers in the United States. The morning started with a vineyard tour, followed by a luncheon and wine tasting. The tastings of Viognier, followed by other Rhone varieties, lasted late into the evening for four of the most devoted audience members. John suggested that the next meeting should include all Rhone varieties to attract many more producers. Mat liked the idea, and the second annual meeting, "Hospice du Rhone," although it was still known as the Viognier Guild, was held at the Alban Vineyards in San Luis Obispo County in 1992. John characterized the attendees as a who's who of future Rhone heavy hitters.

By 1997, the attendees and the scope of the event had grown so rapidly that it was moved to the Paso Robles Event Center at the Mid-State Fairgrounds. Vicki Carroll joined Mat and John to provide the structure and staffing needed to make it a world-class educational organization. Vicki

Vicki Carroll and John Alban greeting guests at a Hospice du Rhone event in Paso Robles. *Courtesy of Hospice du Rhone.*

is the director and president of Hospice du Rhone, a nonprofit educational international organization that attracts attendees from all over the world. In 2010, Hospice du Rhone was invited to present an annual event at Blackberry Farm in Walland, Tennessee. The twentieth anniversary of Hospice du Rhone in 2012 attracted 1,200 people to Paso Robles with over eight hundred wines to enjoy. After a hiatus of four years, Hospice du Rhone resumed its meetings by popular demand in 2016. The new event format will be to schedule meetings every other year.

THE ROBERT HAAS STORY

Robert (Bob) Haas and his partners, Jean-Pierre and Francois Perrin, changed the wine history of Paso Robles when they purchased the land that is now celebrated as Tablas Creek Vineyard. Their first important decision was to become partners in 1987. In 1989 and 1990, the partnership made their second important decision: they decided to make the financial investment to import grapevines. Each clone selected for its high quality was sourced from the famous Châteauneuf-du-Pape in Provence. Each clone was subjected to a minimum of three years in quarantine, coordinated with the United States Department of Agriculture (USDA). After receiving approval, the vineyards at Tablas Creek were planted. In addition to the vines' quality, the advantage of importing the clones was that it provided the vineyard access to the full complement of Rhone varieties.

Tablas Creek Winery introduced a new standard in farming. The Perrin family had been farming organically since the 1950s; however, this practice was not the norm in California viticulture in both the vineyards and university viticultural training. Tablas Creek vineyards were planted organically to produce wines with a sense of place, according to Jason Haas. He described the initial plantings of traditional Rhone varieties, such as Mourvedre, Grenache, Syrah and Viognier. New varieties to America, Grenache Blanc, Counoise and Picpoul Blanc, were also planted. The first vintage of estate wines were produced in 1997. The Chateauneuf tradition is to make blended wines, exemplified by Tablas Creek Rouge and Tablas Creek Blanc. Bob Haas had gradually moved toward the American tradition of labeling wines by varietal. The 2002 Roussanne was the first varietal released at Tablas Creek.

The late Robert Haas shaped the appreciation of fine wine over his long career as a retailer, importer of French wines, wine broker and winemaker. His strategic partnership with the Perrin family established the Rhone varieties in western Paso Robles. *Courtesy of Jason Haas.*

A layer of morning fog covering Tablas Creek Vineyards. *Courtesy of Jason Haas.*

Bob Haas helped develop the California genetic profile for Rhone varieties by establishing a commercial nursery to propagate cuttings for growers. Over two million buds and cuttings have been sold to California growers.

Bob Haas passed away in 2018; he was born in Brooklyn, graduated from Yale and entered the retail wine business as an employee in his father's store in midtown Manhattan. At the age of twenty-seven, he traveled to France to find an agent to import French wines. His previous connection, Raymond Baudoin, had died of a stroke, so Bob decided to take on the challenge of becoming his own agent and working directly with French wineries. This transformed the M. Lehman Liquor Store into a fine wine shop. The store was sold in 1961, which gave Bob the freedom and opportunity to become a wine broker. He moved his business, Vineyard Brands, to Vermont. He met the Perrin family when he started representing the Chateau de Beaucastel, located in Châteauneuf-du-Pape.

Jason Haas moved to Paso Robles in 2002, representing the second generation of the Haas family to do so. Today, he is a partner and oversees

Bob Haas and Jacquez Perrin of Chateau de Beaucastel, shown here in Tablas Creek in 1989, formed an enduring partnership and established a Paso Robles vineyard of Rhone varieties that were imported from France. *Courtesy of Jason Haas.*

Jason Haas, a Tablas Creek Vineyards partner and general manager, with his father, the late Robert Haas, the founder of Tablas Creek Vineyards and renowned importer. *Courtesy of Jason Haas.*

the marketing and business aspects as a general manager at Tablas Creek. Jason, like his father, studied and traveled extensively before joining the business. He was a double major in economics and architecture in college, and he earned a master's degree in archaeology. He spent four years learning about management and marketing at a startup in Washington, D.C. Jason values the continuity of his staff, including winemaker Neil Collins, a native of Bristol, England, who joined Tablas Creek in 1998.

The Rhone Rangers in San Luis Obispo County

The Rhone Rangers are a group of local winemakers who are members of the nonprofit by the same name. Their mission is to promote wines that contain at least 75 percent of the twenty-two Rhone varieties. The name first appeared as a pun on the image of the famous masked man known as the Lone Ranger in early radio and television programs about the mysterious man on a horse who saved others from peril and then disappeared back into the Wild West. The article titled "The Rhone Ranger" in the April 15, 1989 issue of *Wine Spectator* featured founder and winemaker Randall Grahm of Bonny Doon Winery dressed as the Lone Ranger.

The group, first organized in the 1980s, became a nonprofit with a new identity as the Rhone Rangers in the 1990s. Its membership includes wineries in California, Washington, Oregon, Idaho, Michigan and Virginia. The California Rhone movement is centered in Paso Robles, with the San Luis Obispo chapter providing nearly half of its national membership.

The iconic Heart Hill Vineyard, located at Niner Wine Estates in Paso Robles. *Courtesy of Niner Wine Estates.*

Historic Sites to Visit

Eberle Winery
Tablas Creek Vineyard
Hospice du Rhone Events
Verdad & Lindquist Family Wines
Pismo Preserve
Downtown Pismo Beach

10

The Twenty-First Century

New Technology and Trends in Winemaking

GRAPE VARIETIES: NEBBIOLO, SANGIOVESE AND AGLIANICO

Small Brands and Alternate Wines: Dave Caparone and the Noble Italian Varieties

It is said that the next best thing to being in Italy is to sip Nebbiolo crafted by winemaker Dave Caparone. Dave was the first grower to focus on using the legacy of great Italian wines rather than French wines as the gold standard for California winemaking. His focus on premium grape growing led to his search for the best microclimate in which to grow the noble Italian varieties, Sangiovese, Nebbiolo and Aglianico, to proper maturity in San Luis Obispo County. Dave was the first to complete extensive research on these three noble Italian varieties. His father translated articles written in Italian dating back to the early 1930s; in these articles, David discovered recommendations stating that the Nebbiolo should not be grown in California. Dave conducted a six-year study of the microclimates in San Luis Obispo County to select the best location to plant his own vineyard with the three Italian varieties. He was the first to determine that most Italian varieties are planted in the wrong microclimates in California.

Dave was the first grower to select the most suitable clone of Nebbiolo from UC Davis to grow in the Paso Robles AVA to produce a rich, complex wine. In 1980, he planted the first experimental vineyard of Nebbiolo in San

Dave Caparone, the founder of Caparone Winery, in 1979. *Courtesy of Dave Caparone.*

Jazz trombonist and winemaker Dave Caparone is known for producing three extraordinary Italian varietals: Agliano, Nebbiolo and Sangiovese. *Courtesy of Dave Caparone.*

Luis Obispo County and was the first to succeed commercially in producing the Italian varietal Nebbiolo in the United States in 1985.

Dave also planted the first Sangiovese vines in San Luis Obispo County in 1986; the vines were propagated from cuttings brought to Sacramento from the famous Il Poggione Estate in Montalcino in Italy. Dave was the first to research the Aglianico grape and obtain cuttings from both UC Davis and the plant collection of the National Clonal Germplasm Repository, a federal agency that is a part of the Department of Agriculture (USDA). He planted Aglianico in 1987. Caparone Winery was the first in the United States to commercially produce both Italian varietals, Sangiovese in 1986 and Aglianico in 1992.

By the 1990s, Caparone Wines proved that wine growers in San Luis Obispo County were capable of producing premium-quality grapes and world-class wines. The quality of grapes grown and the Caparone wines produced were comparable to their finest Italian counterparts; this has since been recognized by many Italian wine critics in Italy and the United States. The winemaking of Dave Caparone is unique and was developed over the years through the combination of his own research, experience, observation and palate. Dave does not fine, filter, centrifuge, micro-oxygenate or accelerate the winemaking process. His style produces wines that are 100-percent varietal, and they age well.

His noble Italian wines were discovered at a wine tasting by the owners of a small chain of grocery stores known as Trader Joe's in 1984. Dave increased his production to ten thousand cases annually, supplying Trader Joe's with wines under the Caparone label for the next twenty years. The wines developed a loyal following, which continues today.

Caparone Winery, which is family-owned and -operated, dates back to 1979 and is one of the longest-running wine businesses in San Luis Obispo County. Dave's son Marc has worked with him in winemaking and grape growing since childhood. The winery, tasting room, wine club and vineyards are staffed by just two men: Dave and his son Marc.

Custom Crush and the Man Who Filled 190 Million Bottles

As of the 2019 harvest, Tom Myers was recognized as the winemaker who, during the past forty-one years, has filled over 190 million bottles with San

Luis Obispo County wine. How can one man have made enough wine to fill 190 million bottles? The answer: Tom Myers had the expertise, passion and opportunity to produce more than enough wine to fill those bottles. The alchemy and artistry of winemaking continue to fuel Tom's passion for making wine.

Tom developed his expertise after receiving a Christmas gift containing a home winemaking kit from his wife, Kathy. That Christmas was a cold winter day in Michigan, and Tom was intrigued. In 1974, he applied for a federal license from the commissioner of the Bureau of Alcohol, Tobacco and Firearms to become a home winemaker. Tom attended tastings at local wineries, and the more wine he made, the more interested he became in the science of winemaking. Tom applied to UC Davis in California and was accepted into the professional science master's program in viticulture and enology. So, when he was hired as a winemaker, he was the first to have that degree in San Luis Obispo County. Tom is now recognized for his expertise in the science of making wine.

The combination of his background in science and his extraordinary palate ignited his passion for making wine. A great winemaker has a sensitive and acute taste for wines; Tom is described as the winemaker with a Rolodex in his brain that holds the memory of each wine he has ever tasted, analyzed or critiqued. He is known as the problem solver, a resource for every winemaker in the county.

Tom joined the staff of the legendary Estrella River Winery in Paso Robles as an assistant winemaker in 1978; he became the winemaker in 1982. This winery was the first in the county to establish a lab for testing wine with state-of-the-art equipment and technology. The wines Tom made there have served as the prototypes for the trends and styles in the Paso Robles AVA, including Estrella River Cabernet Sauvignon, Castoro Cellars Zinfandel and red blends. Tom is also recognized as one of the great Zinfandel winemakers. "The distinctive and amiable characters of Zinfandel entitle it to rank among the noble varieties of the world. It is our heritage grape," said Tom.

His opportunity to fill 190 million bottles opened up when Tom joined Niels Udsen at Castoro Cellars as the winemaker in 1990. Tom met Niels at Estrella River Winery when Niels joined the staff in 1981. When Tom became the winemaker at Estrella River Winery in 1982, many new wineries were opening around Paso Robles. Most of these wineries were small with little experience in viticulture or enology. Estrella River Winery also produced a surplus of wine grapes from its own vineyards, so the winery offered custom crush services and bottled wine under established labels, including Mission View in San Miguel, Barron and Kolb, Gary Eberle, John Munch and Niels Udsen in Paso Robles. The custom crush business provided wine grape growers and clients with the services of making wine and bottling it with the client's own label. Caparone and Zaca Mesa Wineries brought their own fruit to crush at Estrella River Winery.

Winemakers John Alban of Alban Vineyards, Steve Dooley of Stephen Ross Wine Cellars and Randall Grahm of Bonny Doon have collaborated with Tom to produce their wines. Tom also works with Trader Joe's to make their brand of Central Coast wines. Perhaps the best description of his talents came from local winemakers: "Tom Myers is the awesome winemakers' winemaker."

The First Straw Bale Winery: Clay and Fredericka Thompson

In the early 1970s, Clay Thompson and Fredericka Churchill were in academics, teaching at the University of Michigan. Clay, a recent doctoral degree recipient, was a professor of Old Norse studies, while Fredericka taught German language courses in the same department. In the spring of 1981, Clay was invited to give lectures at UCLA and Berkeley. He and Fredericka decided to drive up the California coast and visit some of the small, artisan wineries that were then being founded. During their third stop at Edna Valley Vineyard, Clay got into a discussion with the cellar crew and began to contemplate leaving academia to start a new career in the wine industry.

When the couple returned to Ann Arbor, Clay called Dick Graff, the co-owner and director of winemaking at Edna Valley Vineyard. After a two-hour interview, Dick hired Clay to be the new "cellar rat" at Edna Valley Vineyard. By mid-August, Clay and Fredericka were married and on their way (by train) to San Luis Obispo. Clay started harvest with the crew in the fall of 1981, just as the final touches were being made to the new Edna Valley production facility and just before the harvest was beginning. During his five-year stint at Edna Valley Vineyard, Clay was able to learn all the aspects of winemaking from the ground up.

The Thompsons' lives and palates were forever changed when they decided to make wine under their own label and specialize in fruity but dry wines in the Alsatian style. Fredericka had spent two years in Germany, just across the border from Alsace, and they were both fond of these unique wines. They traveled to Alsace, hiked the "Wine Road" and discussed winemaking while tasting at many stops along the way. When they returned home in 1983, they founded Claiborne & Churchill Winery and made their first vintage of Dry Riesling and Dry Gewurztraminer in the Edna Valley Vineyard cellar, with grapes purchased from the Niven's Paragon Vineyard.

Throughout the 1980s, the Thompsons continued to expand Claiborne & Churchill, making wines at Edna Valley Vineyard until 1986; they subsequently made wine in an industrial warehouse in San Luis Obispo. In 1990, they were able to purchase six acres in the Edna Valley, making plans for a new winery and tasting room. They wanted something that would fit the rustic, rural landscape, and on the recommendation of a friend, they hired local architect Marilyn Farmer of Habitat Studio, whose focus was on sustainability. Marilyn drew up the plans for an eco-friendly straw bale winery with walls made of stacked bales of rice straw.

In August 1995, Clay and Fredericka broke ground and completed the grading, concrete pad and post-and-beam building. On a sunny Saturday

Clay Thompson, the founder of Claiborne & Churchill, and Jack Niven, the founder of Paragon Vineyards. *Courtesy of Clay Thompson.*

Fredericka, Elizabeth and Clay Thompson celebrate the completion of their winery, Claiborne and Churchill, the first straw bale winery built in California. *Courtesy of the Thompson family.*

in early November, some environmentally minded volunteers and friends gathered to raise the walls of the new winery under the supervision of experienced straw bale contractors. Four hundred bales of rice straw from a farm in the Sacramento Delta were used to build the walls, replacing wood studs, fiberglass insulation and drywall. Three layers of stucco completed the energy-efficient structure. Traditionally, rice straw is burned in the fields, as it doesn't decompose easily. This resistance makes it ideally suitable for building insulation.

The new Claiborne & Churchill Winery building opened for business on January 31, 1996, having the distinction of being not just the first straw bale winery but California's first commercial straw bale building. It has inspired other straw bale wineries and facilities across the globe and has delivered many advantages over the years. It basically recycles a "waste material" and eliminates air pollution from straw burning. The winery requires no heating or cooling because of the insulation value of the straw bale walls, conserving both energy and money.

Don and Gwen Othman: The Bulldog Pup

Before Don Othman started his career as a winemaker, he worked with exotic metals and was an expert in fabricating and welding high-tech metals. He turned down a position in the space industry to move to San Luis Obispo in 1975 with his wife, Gwen. In 1979, they started Bulldog Welding and Manufacturing just prior to the launch of the Central Coast wine revolution. A portable shop was set up on their two-and-a-half-ton 1950 Mack truck, the inspiration for their business name. They supplied stainless steel materials and all the valves, tools and fittings needed for new construction and custom work for wineries in Santa Barbara and San Luis Obispo Counties. There were only a handful of wineries in the area in 1979, and over the next several years, the area's wine industry expanded exponentially.

Don discovered that winemakers faced a challenge when transferring Pinot Noir from the barrel to the bottling tank with a pump. Many of the desirable characteristics, aromas and flavors of the delicate variety, were lost due to the agitation and oxidation pumping caused. To solve this huge issue, Don created a gas-pressure-racking device to gently move the wine out of the barrel using an inert gas, protecting the wine from oxygen and agitation. The first gas-racking wand was fabricated for and tested at Edna Valley Vineyard in 1985. When other winemakers heard how Don's gas-racking wand retained the quality of Pinot Noir from barrel to bottle, requests for the device came pouring in. To meet the demand, the Othmans put the tool into production and named it the "Bulldog Pup." The "Pup" became the standard worldwide and revolutionized how wine, craft beer and spirits were moved in the cellar. It even inspired a new industry-wide term for gas barrel racking: "bulldoggin."

Don and Gwen have always cultivated their land and were home brewers and winemakers. Through Bulldog manufacturing, they worked closely with the best winemakers, whose advice and support helped hone their winemaking skills. In 1995, Gwen and Don embarked on their own commercial winemaking venture and established Kynsi Winery in the Edna Valley. The winery and tasting room are located in a former dairy, which operated on the site in 1965. *Kynsi* means "talon" in Finnish, a nod to the family's heritage and to the barn owls that protect the vineyards and surrounding property. They specialize in Pinot Noir from their Estate Stone Corral Vineyard and also produce an array of wines from renowned cool-climate regions.

Bulldog Pup, a gas-pressure-racking wand created by Don Othman, is used to transfer wine to and from oak barrels without agitation or oxidation. *Courtesy of the Othman family.*

The World of Pinot Noir: Archie McLaren and Brian Talley

In 1996, Brian Talley, commemorating the tenth anniversary of Talley Vineyards, hosted the first livestream dual blind wine tasting event in the United States. Stephen Tanzier, the author of *International Wine Cellar*, hosted several of New York's finest restaurant wine directors to taste the wines in front of a large television screen. At the same time, Brian was hosting a panel of winemakers and a live audience in Arroyo Grande, California. Both of the groups tasted Pinot Noir in an effort to identify the differences in terroir between the Edna Valley, Arroyo Grande Valley, Santa Maria and Santa Ynez AVAs' wines from the Central Coast of California.

Wine enthusiast Archie McLaren attended the dual tasting and recognized the opportunity to build awareness of Pinot Noir with a world-class event in San Luis Obispo County. Brian Talley and Archie McLaren designed the World of Pinot Noir event in 2000 to attract visitors to the Central Coast to taste wine from local producers within Santa Barbara and San Luis Obispo Counties. The inaugural World of Pinot Noir (WOPN) event was held at the Cliffs Hotel in Pismo Beach to educate members of the wine trade, media and consumers about the high-quality wines being produced on the Central Coast of California.

The very best Pinot Noir producers from the Central Coast "came to the table," including Ken Brown (Ken Brown Wines), Kenneth Volk (Wild Horse Winery), Jenny and Dick Doré (Foxen Canyon), Kathy Joseph (Fiddlehead Cellars), Chuck Ortman (Meridian Winery), Frank Ostini (Hitching Post Winery) and Mike Sinor (Domaine Alfred and Sinor-LaVallee). Organizers invited top Burgundy producers, Domaine Leroy and Domaine d'Auvenay from Burgundy, the first year, followed by Domaine de la Romanée Conti in 2001. In subsequent years, World of Pinot hosted producers from major Pinot Noir–producing areas across the globe, including South Africa. The event now takes place at the Ritz-Carlton Bacara Resort in Santa Barbara. The funds raised at the event are donated to the research conducted by professionals in the viticultural programs at Cal Poly University and Hancock College.

The Central Coast Wine Classic and the Rise of Philanthropy

Archie McLaren is the man remembered for creating and sustaining the Central Coast Wine Classic, the top regional wine event, for over three decades. Over the years, this wine event raised over $3.2 million to support philanthropy for public radio station KCBX, the Healing Arts, Performing Arts and Studio Arts in Santa Barbara and San Luis Obispo Counties. He focused the spotlight on the extraordinary wines made in central California and profoundly elevated their quality. He formed the lasting bond between local philanthropy groups and the wine industry; the wineries are major supporters of charities in San Luis Obispo County.

Archie's mission was to introduce the wine lovers and collectors, as well as international wineries, to the local winemakers of the Central Coast. Archie

Archie McLaren and Brian Talley raising money for the Fund for Vineyard and Farm Workers at the Central Coast Wine Classic in 2017. *Courtesy of the Wine History Project of San Luis Obispo County.*

believed that the wines of the Central Coast could hold their own against any wines in the world. He brought local wines and winemakers to the attention of the international collectors, wine critics and international winemakers by inviting them to wine auctions, barrel tastings and elegant dinners in local venues from Hearst Castle to local wine cellars. Archie believed in food and wine education for those attending and accomplished this by offering food demonstrations and educational wine seminars at the annual Wine Classic. The wine auction items included lifestyle events, trips abroad, rare wine dinners and trips with Archie as the host. By offering these auction items, he expanded the annual Wine Classic experience and party over the following twelve months. No one else was creating this type of wine auction experience.

The annual Wine Classic events began in 1985 and continued for thirty-three years. Usually, the event was celebrated for three or four days in San Luis Obispo County and included barrel tastings with local winemakers; dinners at Hearst Castle, Paso Robles and Edna Valley wineries; educational symposiums; and all-day-long live auctions. This auction was the major fundraiser of the event and was followed by the famous "Vintage Dinner" with classic Cuvée tastings.

Champagne brunches were celebrated on Sunday mornings. Sunday afternoon concluded with a free wine tasting for the public with over one hundred winemakers showcasing local wines. McLaren's culinary friendships and patrons included chefs Paul Prudhomme, Susan Spicer, Emeril Lagasse, Wolfgang Puck, Gary Danko, Jacques Pepin, Christopher Eme, Laurent Quenioux, James Sly, Ian McPhee, Michael Hutchings, James Siao and famous cookbook author Julia Child. Archie McLaren was a master at marketing and building a network of relationships to educate and elevate the California lifestyle and promote California wines. His grace and charm made him unforgettable. The final Central Coast Wine Classic was held in August 2017. Archie died of cancer at his home in Avila in February 2018.

Historic Sites to Visit

Caparone Winery
Claiborne and Churchill Winery
Kynsi Winery
Sinor-LaVallee Wines
World of Pinot Noir event in Santa Barbara
San Luis Obispo Coast Wine's Harvest on the Coast Festival

Appendix

Information on San Luis Obispo County Wineries and Historic Sites

Chapter 1

Mission San Luis Obispo de Tolosa
751 Palm Street
San Luis Obispo, CA 93401
www.missionsanluisobispo.org

Mission San Miguel
775 Mission Street
San Miguel, CA 93451
www.missionsanmiguel.org

Point Luis Lighthouse
1 Lighthouse Road
Avila Beach, CA 93424
www.pointsanluislighthouse.org

Morro Rock
www.morrobay.org

Piedras Blancas Lighthouse
15950 Cabrillo Highway
San Simeon, CA 93452
www.piedrasblancas.org

California Mid State Fairgrounds
2198 Riverside Avenue
Paso Robles, CA 93446
www.midstatefair.com

Chapter 2

Dallidet Adobe and Gardens in San Luis Obispo
1185 Pacific Street
San Luis Obispo, CA 93401
www.historycenterslo.org

Cayucos Morro Bay Cemetery
2451 Ocean Boulevard
Cayucos, CA 93430
www.cayucosmorrobaycemetery.com

Old Mission Catholic Cemetery
101 Bridge Street
San Luis Obispo, CA 93401
uscemeteryproj.com

South County Historical Society
126 South Mason Street
Arroyo Grande, CA 93420
www.southcountyhistory.org

Avila Beach and the Bob Jones Trail
www.visitavilabeach.com/bob-jones-trail

The Dana Adobe
671 South Oakglen Avenue
Nipomo, CA 93444
www.danaadobe.org

The Village of Arroyo Grande
www.visitarroyogrande.org

Lake Lopez
6820 Lopez Drive
Arroyo Grande, CA 93420
www.lopezlakemarina.com

Pozo Saloon
90 Pozo Road
Santa Margarita, CA 93453
www.pozosaloon.com

Chapter 3

Epoch Estate Wines
7505 York Mountain Road
Templeton, CA 93465
www.epochwines.com

San Luis Obispo Railroad Museum
1940 Santa Barbara Avenue
San Luis Obispo, CA 93401
www.slorrm.com

The Rural Community of Shandon

Paso Robles Historical Society
800 Twelfth Street
Paso Robles, CA 93446
www.pasorobleshistoricalsociety.org

Paso Robles Pioneer Museum
2010 Riverside Avenue
Paso Robles, CA 93446
www.pasoroblespioneermuseum.org

Steinbeck Vineyards & Winery
5940 Union Road
Paso Robles, CA 93446
www.steinbeckwines.com

Chapter 4

J Dusi Winery
1401 CA-46
Paso Robles, CA 93446
www.jdusiwines.com

Rotta Winery
250 Winery Road
Templeton, CA 93465
www.rottawinery.com

Turley Wine Cellars
2900 Vineyard Drive
Templeton, CA 93465
www.turleywinecellars.com

ZinAlley
3730 CA-46
Templeton, CA 93465
www.zinalley.com

Scenic Highway 46, west from Highway 101 to the Pacific Ocean
www.travelpaso.com

Camp Roberts
San Miguel, CA 93451
www.calguard.ca.gov

Chapter 5

Paso Robles Inn
1103 Spring Street
Paso Robles, CA 93446
www.pasoroblesinn.com

Downtown Paso Robles
www.pasoroblesdowntown.org

Glunz Winery
8331 CA-46
Paso Robles, CA 93446
www.glunzfamilywinery.com

Madonna Inn
100 Madonna Road
San Luis Obispo, CA 93405
www.madonnainn.com

Thursday night Farmers Market in downtown San Luis Obispo
www.downtownslo.org
www.slocountyfarmers.org

Hearst Castle
750 Hearst Castle Road
San Simeon, CA 93452
www.hearstcastle.org

Sebastian General Store and Hearst Ranch Winery Bar
442 Slo San Simeon Road
San Simeon, CA 93452

Piedras Blancas Light Station
15950 Cabrillo Highway
San Simeon, CA 93452
www.piedrasblancas.org

Elephant Seals
Piedras Blancas Rookery
Elephant seals viewing area
Vista Point
San Simeon, CA 93452
www.parks.ca.gov

Carrizo Plain National Monument
17495 Soda Lake Road
Santa Margarita, CA 93453
www.blm.gov/visit/carrizo-plain-national-monument

Chapter 6

Adelaida Vineyards
5805 Adelaida Road
Paso Robles, CA 93446
www.adelaida.com

Le Cuvier Winery
3333 Vine Hill Lane
Paso Robles, CA 93446
www.lcwine.com

Windward Vineyard
1380 Live Oak Road
Paso Robles, CA 93446
www.windwardvineyard.com

Downtown Cambria and Moonstone Beach
www.visitcambriaca.com

DAOU Vineyards
2777 Hidden Mountain Road
Paso Robles, CA 93446
www.daouvineyards.com

Chapter 7

Edna Valley Vineyard
2585 Biddle Ranch Road
San Luis Obispo, CA 93401
www.ednavalleyvineyard.com

Saucelito Canyon Winery
3080 Biddle Ranch Road
San Luis Obispo, CA 93401
www.saucelitocanyon.com

Old Edna Townsite
1653 Old Price Canyon Road
San Luis Obispo, CA 93401
www.highway1discoveryroute.com/activities/old-edna-townsite/

Biddle Ranch and Corbett Canyon Road

Chamisal Vineyards
7525 Orcutt Road
San Luis Obispo, CA 93401
www.chamisalvineyards.com

Center of Effort Wine
2195 Corbett Canyon Road
Arroyo Grande, CA 93420
www.centerofeffortwine.com

Baileyana, Tangent & True Myth Tasting Room and the Historic Schoolhouse
5828 Orcutt Road
San Luis Obispo, CA 93401
www.nivenfamilywines.com

Wolff Vineyards
6238 Orcutt Road
San Luis Obispo, CA 93401
www.wolffvineyards.com

Chapter 8

Paso Robles Wine Festival
www.pasowine.com

JUSTIN Vineyards and Winery
11680 Chimney Rock Road
Paso Robles, CA 93446
www.justinwine.com

Peachy Canyon Winery
1480 North Bethel Road
Templeton, CA 93465
www.peachycanyon.com

Victor Hugo Winery
2850 El Pomar Drive
Templeton, CA 93465
www.victorhugowinery.com

Kenneth Volk Vineyards
5230 Tepusquet Road
Santa Maria, CA 93454
www.volkwines.com

Vintage Paso Zinfandel Weekend
www.pasowine.com

Talley Vineyards
3031 Lopez Drive
Arroyo Grande, CA 93420
www.talleyvineyards.com

Stephen Ross Wine Cellars
178 Suburban Road
San Luis Obispo, CA 93401
www.stephenrosswine.com

Laetitia Vineyard & Winery
453 Laetitia Vineyard Drive
Arroyo Grande, CA 93420
www.laetitiawine.com

Oceano Dunes Natural Preserve
928 Pacific Boulevard
Oceano, CA 93445
ohv.parks.ca.gov

Chapter 9

Eberle Winery
3810 CA-46
Paso Robles, CA 93446
www.eberlewinery.com
Tablas Creek Vineyard
9339 Adelaida Road
Paso Robles, CA 93446
www.tablascreek.com

Hospice du Rhone
www.hospicedurhone.org

Verdad & Lindquist Family Wines
130 West Branch Street
Arroyo Grande, CA 93420
www.verdadwine.com

Pismo Preserve and downtown Pismo Beach
www.lcslo.org/pismopreserve
www.experiencepismobeach.com

Chapter 10

Caparone Winery
2280 San Marcos Road
Paso Robles, CA 93446
www.caparone.com

Claiborne & Churchill Winery
2649 Carpenter Canyon Road
San Luis Obispo, CA 93401
www.claiborneandchurchill.com

Kynsi Winery
2212 Corbett Canyon Road
Arroyo Grande, CA 93420
www.kynsi.com

Sinor-LaVallee Wines
550 First Street
Avila Beach, CA 93424
www.sinorlavallee.com

World of Pinot Noir event in Santa Barbara
www.worldofpinotnoir.com

SLO Coast Wine's Harvest on the Coast Festival
www.slocoastwine.com

Bibliography

Angel, Myron. *The History of San Luis Obispo County, California*. Oakland, CA: Thompson and West, 1883.

Ausmus, William A. *Wines and Wineries of the Central Coast*. Berkeley: University of California Press, 2008.

Bonne, Jon. *The New California Wine: A Guide to the Producers and Wines Behind a Revolution in Taste*. Berkeley, CA: Ten Speed Press, 2013.

Briscoe, John. *Crush: The Triumph of California Wine*. Reno: University of Nevada Press, 2018.

Cinotto, Simone. *Soft Soil, Black Grapes: The Birth of Italian Winemaking in California*. New York: New York University Press, 2012.

Comiskey, Patrick J. *American Rhone: How Maverick Winemakers Changed the Way Americans Drink*. Oakland: University of California Press, 2016.

Franks, Janet Penn. *San Luis Obispo County Wineries*. San Luis Obispo, CA: Central Coast Press, 2005.

Haeger, John Winthrop. *Pacific Pinot Noir: A Comprehensive Winery Guide of Consumers and Connoisseurs*. Berkeley: University of California Press, 2008.

Hodgins, Paul, and Julia Perez. *The Winemakers of Paso Robles*. Los Angeles: Winemakers Series, 2017.

Jones, Idwal. *Vines in the Sun: A Journey Through California Vineyards*. New York: William Morrow and Company, 1949.

Leon, Vicki. *California Wineries: San Luis Obispo, Santa Barbara, and Ventura*. San Luis Obispo, CA: Blake Publishing, 1986.

Lukacs, Paul. *American Vintage: The Rise of American Wine*. New York: W.W. Norton and Company, 2000.

Morrison, Annie L., and John H. Haydon. *Pioneers of San Luis Obispo County & Environs*. San Miguel, CA: Friends of the Adobes Inc., 2002.

Pinney, Thomas. *The City of Vines: A History of Wine in Los Angeles*. Berkeley, CA: Heyday, 2017.

———. *A History of Wine in America: From the Beginnings to Prohibition*. Vol. 1. Berkeley: University of California Press, 2007.

Rice, Thomas J., and Tracy G. Cervellone. *C.W.E. Paso Robles: An American Terroir*. Paso Robles, CA: Self-published, 2007.

Roberts, Larry, and Carol Manning. *Vineyards on the Mission Trail: The Wineland of Santa Barbara and San Luis Obispo Counties*. Santa Maria: California Central Coast Wine Growers Association, 1981.

Sullivan, Charles L. *A Companion to California Wines*. Berkeley: University of California Press, 1998.

About the Authors

Libbie Agran, Founder and Historian of the Wine History Project of San Luis Obispo County

Libbie Agran is the founder of the Wine History Project of San Luis Obispo County. This project archives and preserves the fascinating history of two hundred years of wine growing on land where over 150 varieties of grapes were planted and world-class wines were produced. This project brings the area's history to the public through exhibits in vineyards, wineries, historic houses, museums and along hiking trails. The website, historic timeline, lectures, films and tastings are hosted by the Wine History Project. Libbie grew up in Southern California among orange trees and grapevines. She came of age during the California wine revolution and has been drinking Zinfandel and Pinot Noir on the Central Coast ever since. She collects oral histories and writes the "Legends" and articles on the website. She has produced two documentaries about famous local winemakers and grape growers. Libbie is a graduate of UCLA with a bachelor's and master's degree. This is her second book, but it is the first wine history of San Luis Obispo County. Libbie also wrote *Archie McLaren: The Journey from Memphis Blues to the Central Coast Wine Revolution*, which was published by the Wine History Project of San Luis Obispo County.

Heather Muran, Historian of the Wine History Project of San Luis Obispo County

Heather Muran worked as a historian in 2019 and 2020 for the Wine History Project of San Luis Obispo County. She developed a passion for the local wine industry while working weekends as a tasting room attendant. She holds a degree in journalism from San Diego State University and has spent the last decade promoting San Luis Obispo's coastal wines to traders, the media and consumers across the globe.

Heather collaborated with growers and vintners to collect and archive viticultural history for the Wine History Project. She enjoys living "the SLO life" with her family, exploring the great outdoors and beauty of San Luis Obispo County through surfing, hiking, sailing, cooking and pairing cuisine with local wines.